iPhone App Design for Entrepreneurs

Find Success on the App Store without Coding

Megan Holstein

Apress®

iPhone App Design for Entrepreneurs: Find Success on the App Store without Coding

Megan Holstein
Dublin, OH, USA

ISBN-13 (pbk): 978-1-4842-4284-1 ISBN-13 (electronic): 978-1-4842-4285-8
https://doi.org/10.1007/978-1-4842-4285-8

Library of Congress Control Number: 2019936304

Managing Director, Apress Media LLC: Welmoed Spahr
Acquisitions Editor: Aaron Black
Development Editor: James Markham
Coordinating Editor: Jessica Vakili

Cover designed by eStudioCalamar

Cover image designed by Freepik (www.freepik.com)

Distributed to the book trade worldwide by Springer Science+Business Media New York, 233 Spring Street, 6th Floor, New York, NY 10013. Phone 1-800-SPRINGER, fax (201) 348-4505, e-mail orders-ny@springer-sbm.com, or visit www.springeronline.com. Apress Media, LLC is a California LLC and the sole member (owner) is Springer Science + Business Media Finance Inc (SSBM Finance Inc). SSBM Finance Inc is a Delaware corporation.

For information on translations, please e-mail rights@apress.com, or visit http://www.apress.com/rights-permissions.

Apress titles may be purchased in bulk for academic, corporate, or promotional use. eBook versions and licenses are also available for most titles. For more information, reference our Print and eBook Bulk Sales web page at http://www.apress.com/bulk-sales.

Any source code or other supplementary material referenced by the author in this book is available to readers on GitHub via the book's product page, located at www.apress.com/978-1-4842-4284-1. For more detailed information, please visit http://www.apress.com/source-code.

Printed on acid-free paper

A huge thank you to Valerie and Michael Holstein, my parents, for believing that a fifteen year old could start a company. You are the parents who said "yes."

"This one step—choosing a goal and sticking to it—changes everything."

—Scott Reed

Table of Contents

About the Author

Megan E. Holstein is an entrepreneur and fiction and nonfiction author. Her youngest brother has severe autism. This led her to start a business at 15 making apps for autistic children called Pufferfish Software. It went on to be a successful medical app company, serving tens of thousands of autistic patients. She and her work with Pufferfish have been featured in *Inc.* magazine and *Business First.* She won the Global High School Entrepreneur award from the Entrepreneur's. See more about Megan at `www.meganeholstein.com`.

Introduction

This book is for people who aren't geeks. To read this book, you don't need to know how to program, do graphic design, or possess any especially technical skillsets at all. This book is full of information that will help anyone make their own app, not just anyone with coding experience.

To get started, you should know the basics of how computers work, including how they store and transfer data, and the basics of how programs work. If you do not know these things, research the basics of how computer hardware and software works before returning to this book. Smartphones work the same as small computers, although they have vastly different interfaces and uses.

If you don't like programming, you can still make your own app.

This book won't teach you how to code, because an app is more than code. Apps are small companies, and therefore require marketing, customer relations, product design, and all sorts of things in addition to programming. It is for this reason that expert programmers don't always create good apps, and good apps are not always created by expert programmers. Being a great programmer doesn't automatically make you a great small business owner.

Programming the app is only one part of one of building an app; to place the entire book's emphasis on just that would be to neglect other important parts of a business.

If you plan and make intelligent decisions when it comes to your small app business, you will find success on the App Store. It would be a lie to say that this book has all of the "secrets to success" for the App Store, but you can count on success in the App Store if you work hard and plan. The App Store is not a lottery. Apps are not featured at random, and it is not random whether or not your app becomes successful.

There is an element of luck and timing to success, but working hard and creating a good app play a much bigger part. You cannot just whip up an app, throw it on the App Store, and watch the dollars roll in—nor will you slave away for years at an app, only for it to make no money. You will get what you deserve.

Do You Really Need an App?

Everyone these days has an idea for an app, from the garbage collector to the mailman, to your boss. There is always an app for that, too, from a digital girlfriend app to an app that helps you manage multiple girlfriends.

It is popular to think that you or your business needs an app on the App Store. However, this might just not be true. An app is a fantastic way to provide a service, but an app is not an end in and of itself. Apps are just one way of getting things done on a computer.

To illustrate this, we will take a look at the relationship between desktop applications and websites in the early days of the Internet—a good model of the relationship between apps and the Internet today.

Let's hearken back to 1987, the release of the color display Macintosh II.

These were the days when everyone felt "you just had to be on the computer," just as people feel "you just have to have an app" today. The Macintosh II computer had dedicated applications for everything, and you only went on the Internet to access basic text and information. Heavy computing weight was pulled by standalone applications such as Word and Outlook, and the Internet was just for finding information or for connections inside the dedicated applications.

This resembles the mobile computing experience now. You use your phone browser occasionally, but most of your time is spent inside an app. Apps are what cost your phone the most processing power, and the Internet is utilized inside other apps more than it is by the browser.

But things did not stay that way for computers. As Internet technology advanced, more and more functionality moved onto the browser. Dedicated games and applications have all but disappeared from your desktop, and all you're left with are a couple of large and heavyweight applications such as text, photo and video editors, and in-depth games.

Most of what you do has moved onto the browser—you connect with your friends in the browser, your cloud storage is in your browser, your homework is done in the browser, you check your email in the browser... all things that used to be restricted to dedicated applications.

What Does This Mean for Apps on Smartphones?

It means that things are likely moving the same way for the smartphone.

Simple applications are going to move into the browser as better Internet technologies become more popular and supported, and only big and heavyweight functions of smartphones are going to remain dedicated inside their own applications.

So no, apps are not the end-all-be-all of the smartphone. In fact, if the trend follows history, the app will be largely phased out. We're still at an early point in this evolution, where content-consuming apps remain as standalone applications on smartphones. Because of this, lots of apps do still belong as apps. Some common app ideas used to be appropriate as apps, but are not so anymore, like:

- Apps for a single content-producing website
- Apps for a single ecommerce store
- Apps for a single brand

What all of these ideas have in common is that they're ideas for single content providers. Content aggregators make sense as standalone apps, as a one-stop-ship destination for the user. (This would be Amazon, Netflix, or the iTunes Store.) However, single-producer websites are better off as responsive websites instead of apps within themselves, because the users aren't going to be visiting these websites very often.

You don't want to force your user to download an app just to visit a website a couple of times a year.

That doesn't mean you should just forfeit the smartphone market, though. If your app seems like something the user wouldn't use all that often, consider if it's better off as a responsive website. A responsive website is a website that works just as well on a phone as a native app would. Your idea is probably better off as a responsive website if:

- The user will not be visiting your app more than once every fortnight
- Your app is going to be an outlet for a single provider
- Your app will have a limited amount of content
- Your app will have static content (such as an encyclopedia)

If your idea fulfills these points and you're still set on making an app, make sure you've thought through why making an app is better than just making your website work on phones. Your app must add so much value that users would prefer it over a mobile website, immediately.

Unnecessary Expenditures

You want to be on guard against unnecessary spending on your app-making journey. It can be tempting to spend a lot of money when you're trying to make an app, as you find things that will make you more legitimate. These can range from inexpensive things like URLs and business cards, to expensive things like web design services and consultation.

Don't let concerns about being official, legitimate, or anything else convince you to spend money where you don't otherwise need to. For the first version of an app, you're only looking to prove that the concept of your app works. If it doesn't, you don't want to have lost a ton of money in the process.

Save expenditures that make you seem legitimate until after your first version is released; until after you know you don't have a dud on your hands.

Do You Need an NDA?

We're going to take a moment and specifically address this myth before we get started, because believing in it will seriously impede your progress throughout this book.

The myth is that you need an NDA to protect your app idea from Zuckerberg-style intellectual property theft. It's not true.

You do not need an NDA to protect your app idea.

NDAs do have their time and place. For instance, if you're Apple Inc. and are striking a deal with AT&T for exclusive service on the new-and-still-secret iPhone, you want an NDA. If AT&T were to leak that Apple was making a smartphone, it would have ruined Apple's plans and changed the smartphone market for years.

But, you are not Apple Inc. You are not striking a high-stakes deal with two billion dollar companies, which if revealed could ravage tens of thousands of families. If your idea is shared, you won't drop dozens of points in the New York Stock Exchange. You have nothing to lose, because you have nothing at all.

Furthermore, Apple had a real and concrete product, not an idea, when they were pursuing a deal with AT&T. The AT&T NDAs were not protecting Apple executive's *ideas* for an iPhone, they were protecting *the iPhone*. All you have is an idea.

An idea that, likely, someone has had before. Probably many people have had before. When starting to make an app, all anyone has is an idea.

People Don't Respect NDAs in App Development

Quite apart from the needlessness of an NDA, it's counterproductive to even request one. Many industry professionals will outright refuse to sign one, and you don't want to lose the valuable resources of money and mentorship over an NDA.

One of the signs of a new, inexperienced app developer is an insistence upon an NDA.

Whenever someone says, "I have a great idea, but I can't talk about it without an NDA," those with experience don't think, "Ooh, I wonder how great their idea is! I'm so curious!" They think, "Oh, great, not another one." People who demand NDAs become repellant to the successful when it comes to the field of mobile apps.

If you do demand an NDA, your company probably won't get very large anyhow.

Lastly and most importantly, the *only way to validate your idea* (make sure it's even worth the effort) is by talking to a lot of people. You will have to tell people about your idea to do so. There is an *extremely small* chance that one of these people will steal your idea, but that segues into the next point.

It's all about execution anyway. Without execution, your idea is literally worth nothing. The only value an idea has is because it is being executed on; your app idea, if you decided not to make it a reality, would instantly become worthless unless someone else tried to make it a reality.

Ideas are so worthless on their own that Seth Godin put 999 of them on his blog for free. Look through those ideas and you'll notice that some of them are actually pretty great. If ideas were worth so much, Seth Godin would have sold them or protected them instead.

If you still feel that you need to request an NDA, make sure your reason for doing so is unique and compelling. Be able to defend your use of an NDA when you inevitably encounter resistance.

PART I

Design

“It is not simplicity on this side of complexity I am concerned with, but simplicity on the other side of complexity.”

—Oliver Wendell Holmes

CHAPTER 1

Refining Your Idea

A great idea with poor execution is a poor app, but a poor idea with excellent execution is a great app. This means that it matters far less what your app idea is, and far more that it's a well made app.

This is true, but it's not the whole truth. Anyone who has failed at a startup can verify that no matter how well you execute a poor app idea, you can never be very successful, because part of what makes a poor app idea poor is that nobody wants to download the app. There is no market demand for the app.

So yes, having a good idea is the key to success. Having a refined and thought-out idea is part of what makes an app a success. If start making the first thing that pops into your head, you'll have no idea whether or not it will sell. You need to assess and adjust your idea for success now, while it's still free and easy to do so. If you wait until later, it will be exceedingly more expensive to adjust your idea, like turning around on a road trip after you're already halfway to the destination.

This isn't just because you need a winning idea for a winning app; with over one million apps on the App Store, your app idea is competing against hundreds of other apps based on the exact same idea. The humble task manager has thousands of iterations on the App Store, all competing on execution and not on idea. If you make a task manager with just a slightly better idea (such as group collaborating on tasks for school projects), you'd already have a huge leg up on the competition. Every category is similarly crowded on the App Store.

M. Holstein, *iPhone App Design for Entrepreneurs*,
https://doi.org/10.1007/978-1-4842-4285-8_1

Contrary to popular opinion, you don't just magically "have" a better idea. It doesn't happen in a brilliant flash of genius. Having a better idea involves taking a good-enough idea and making it better through hard work. This requires a significant time investment in developing your app idea. Analyze your idea by checking it against the following idea-refinement guidelines.

Make it Small in Scope

One easy point to remember while refining your app idea is that it should be small in scope. This means that it should do only a handful of things and it should do them really, really well. Think about all your favorite apps—Snapchat, the app that sends self-destructing photos, Facebook, the app that lets you post messages, pictures, and comments to people or group walls, Twitter, the app that has 140 character messages, 2048, a game with exactly four controls where you slide tiles around, and so on—the examples are endless. These apps only do a few things, but they do those few things well.

Tip Smartphones are used in small bites, in moments and seconds, and the purpose of your app should reflect this.

An iPhone game should resemble a Flash game, not in graphics but in gameplay (quick play under 30 minutes, during class time or while the boss isn't looking). A productivity app should get you to the screen you need and get you out quickly. A utility app should be even more streamlined—most utility apps have one or two screens at most.

Even the most in-depth apps, like document editing apps, won't be used for more than a half hour or so at a time. Users aren't powering through and writing the great American novel on their iPhones.

At best, they're capturing ideas on their iPads. Any long times spent in front of an iPad are done consuming content (watching TV shows and movies, or studying for class).

With so little time to make an impression, your app needs to make a good one, and make it quickly. You can't do this with an app that has more functions than a Swiss Army Knife.

Tip Your app needs to be simple, new, and engaging, the first time.

Cool Features ≠ a Good App

Another easy guideline for app-idea refinement is that its features should be relevant to the app. Cool features do not make an app good.

When the iPhone first came out, using the accelerometer feature was very in vogue. Apps that became famous early, like iBeer (see Figure 1-1), had ideas that used the accelerometer in an engaging and simple way.

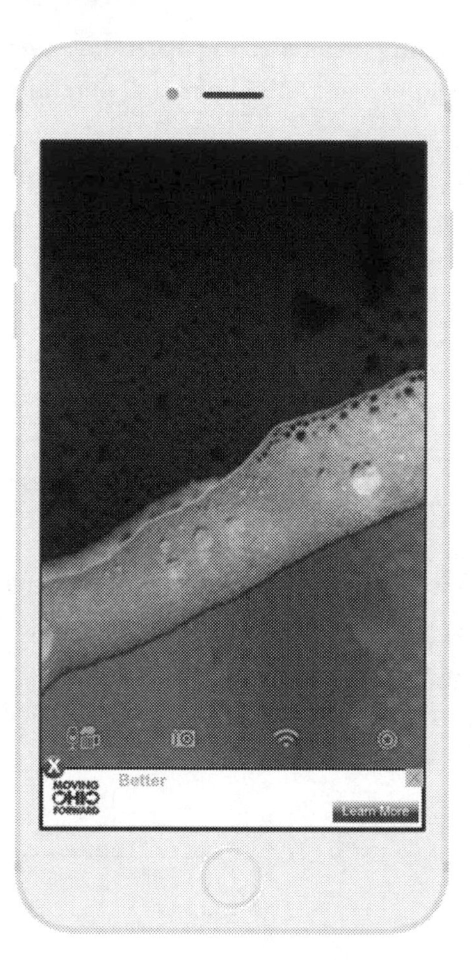

Figure 1-1. *The iBeer App*

Apps that did not become famous put the accelerometer feature before the app's purpose. One example were apps that would exchange business cards simply by bumping phones or putting them near each other. These apps did not get nearly as big as anticipated, because they were classic examples of putting the feature before the purpose.

If you've thought of some feature that would be really cool on iOS, that's great—write it down and save it for an appropriate app. While a feature can make an app cool, new, and different, a feature isn't a reason to make an app in the first place.

With new apps and cool features coming out every day, it can be easy to get wrapped into feature hype. But it's important that you're able to differentiate between a feature and a benefit.

Take the Mailbox app (see Figure 1-2), an email inbox management app available for iOS and Android. It attracted a lot of media attention for having a huge waitlist, and subsequently being acquired by Dropbox.

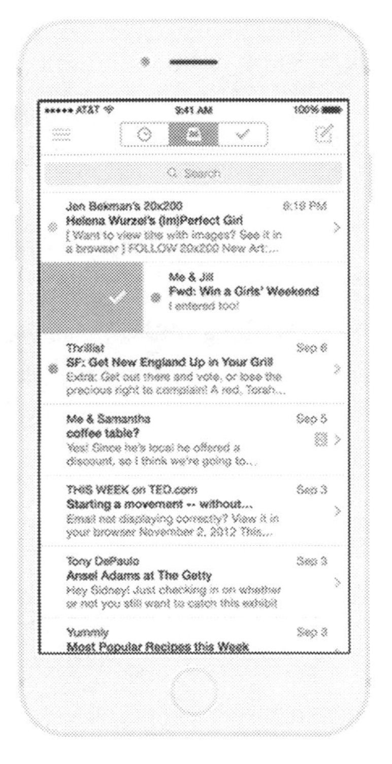

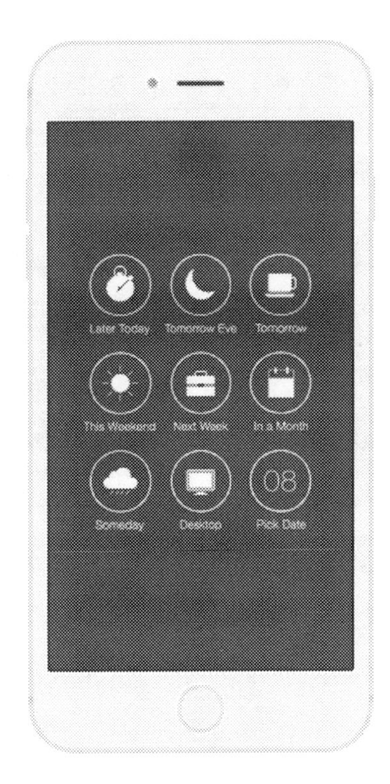

Figure 1-2. *The Mailbox App*

You might say "But Mailbox is cool because it has different features for email!" No, the idea of snoozing an alert or leaving it until later is not a new feature. Our smartphones have been capable of this for a long time.

Mailbox is different because it applied these features to a new core concept, which was "treating email like a to-do list".

As cool as some iOS features are, at the end of the day, apps are purpose-driven. An app's feature has to do something in the context of the function of the app. If it doesn't help the user have a better and smoother experience, it should left out.

The importance of this cannot be overemphasized. Differentiating on feature set won't win you the fight, but differentiating on core experience will.

Tip An app has to have a good core function and a good experience to be successful in the market.

The Apple MacBook is a great example of this. On paper, it has fewer features and hardware than any of its computers and is astronomically more expensive. By every numerical measurement, it's a worse computer than almost any PC. Yet, it's a top seller because it provides a delightful experience.

Providing a delightful experience is what will win you the app market, and providing a delightful experience begins with a refined core app idea.

Compare with Your Competition

Every app has competition, even yours. Competition comes in two flavors:

- *Direct competition:* Direct competition is competition that competes directly with you. They solve the same problem you do, in the same way you do. For instance, Honda and Ford are in direct competition with each other in the automobile industry.
- *Indirect competition:* Indirect competition is competition that solves the same problem you do, but not in the same way.

Taxi cab companies are indirect competitors of Honda and Ford, because all three companies solve the same problem (transportation of passengers), but taxi companies do so in a different way.

Your app may not have any direct competition, but it will always have indirect competition. This competition can be something as simple as a paper and pencil list, or something from another industry entirely. Whatever your indirect competition is, you have it.

What You're Really Saying

When you say you don't have any competition, you're really saying one of these things:

- "We have competitors and don't know it." This is not inspiring. You simply cannot go into business without knowing how your product is better than your competitors, direct or indirect. Find your competition and pin down how you are better.
- "We sell a substitution product." For instance, you sell margarine and all your competitors are butter companies. That's fine, but you still must make sure it's obvious why someone should buy your margarine, and not their butter. Don't rely on people to figure out why they should buy your product on their own.
- "We have no competitors, because nobody wants an app that does this." If your app is unique and you have no competitors, that means your app probably doesn't solve a painful enough problem for your users. This circumstance is unlikely. What is far more likely is that you have competition that's not good at marketing, or you are in a niche category.

- "We have no competitors because we have something awesome people don't know they want yet." It cannot be overstressed how unlikely this is. And even if this is the case, you will have competitors almost the second your app launches. What's most likely is that your idea is not as unique as you think.

Finding Your Competition

If you don't know who your competition is, go out there and find your competition. You can start this search process by answering a couple questions about your users:

- What are your users currently doing to overcome the problem they're having?
- If the users didn't have your solution, what would they do?

Tip Find your competition and pin down why your app is better.

People who lack direct competition have a marketing challenge ahead of them; they need to explain to users why their substitution app is better than the app they're already using. It's easy to convince people to switch from a car to a better car, but not so easy to convince them to switch to a taxi.

On the other hand, in the war between two directly competing apps, users will download the app that's been reviewed more and is more popular. So while it's easier to convince people why they need this solution, it's harder to convince them why they need *your* version of the solution. This is due only to the fact that your direct competition is higher up on the charts and easier to find, because they were there before you.

Because other apps (of any kind) will be higher up on the charts and more reviewed, if there isn't a clear reason to choose your app, users won't. There needs to be a very real and very obvious reason why your app is better, a reason that isn't just tied to the ratings and chart position.

Furthermore, what is obviously better to you isn't obvious. If it's not a short sentence and immediately obvious after seeing just one screenshot of the app, it isn't obvious enough. This is all the investigating a user might be doing, and you need to impress them in that brief amount of time.

Differentiating your app is quite the challenge, but the field is competitive with 1.3 million apps on the App Store. More apps are added every day, and the store is only getting more competitive.

Execution is Everything

Execution is everything. You can create an app that makes your iPhone print money, but if users can't find the Print button, folks won't use the app.

Note Whatever your app does, it has to do it simply and well.

It doesn't matter how many features your app has, or how fancy it is. Consider OmniFocus; it is considered the biggest and most robust task management app on the App Store. It syncs with just about everything, reminds you of to-dos based on time and location, and you can attach links and videos within the app.

Plenty of people like OmniFocus, but just as many people like Clear. Clear is also a to-do app, and it is very basic—you pull down to create item, swipe to complete, pinch to see lists, and then expand again to see the items in that list. It doesn't have nearly as many features as OmniFocus, but has enjoyed a lot of success—about 2,000,000 downloads' worth.

This is evidence of the fact that a well-executed idea can lead to success on a bigger and grander scale than features, any day.

Make sure that not only is your app idea small in scope, but that what it does do is executed in a way that is simpler than anything else available.

Make it Easy to Explain

Another reason iBeer was so successful was because it was exceedingly easy to explain. "It's like beer on your iPhone!" Temple Run is similarly easy to explain. "You've got to avoid the monkeys and not fall off!" Or 2048, another game that went viral recently. "You've got to combine numbers to get to 2048!"

Things that are easily explained easily go viral, because all it takes is one sentence exchanged from one friend to another for a download to happen. By contrast, anything that is difficult to explain will not go viral, no matter how great it is, because people won't invest the time necessary to understand it.

A good way to measure how easy your app is to explain is by handing your iPhone app to someone and asking them to tell their friend about it. Don't provide them any prompting or help; you don't want to provide them with good ways to explain your app or good things to say, because you want to see what a random person out in the world would say to explain it.

If your app idea can't be explained in one sentence by someone else, you need to refine it further. Test different one-liners for your app over and over with people; once your app is easily explained, you're golden.

Always Get Feedback

One of the best ways to measure your app idea is to get feedback. Feedback is important throughout the app-making process because it is a reality check as to what your users want.

Tip Getting feedback early and often is one of the keys to creating an app that people love.

The following sections discuss some ways you can get personal feedback on your idea.

Ask Friends and Family for Feedback

This is something you can do right now; stop reading this and type up an email to a close friend with your app idea and ask them what they think of it. Make sure it is someone who is comfortable with providing you honest criticism, and won't just tell you it's good to encourage you. If your idea doesn't pass this initial test, modify it before moving on to other methods of validation.

Go to Local Events for Your Target Audience

These can be found through the website Meetup, especially in the tech section. Attend some of these app and startup meetings. Tell people about your app idea and solicit feedback from them. Connect to people you meet on LinkedIn and grow your personal network. A personal network is one of those things that you don't need until you need it—and then you really need it.

Tip You've got your user feedback on your idea now—be careful not to put too much weight on their feedback.

Interpreting User Feedback

Knowing what your user wants and needs is a critical part in designing a successful app, but you need to be careful not to just mindlessly chase the features the users ask for. User feedback can help center your design, but you need to always keep the core purpose of your app in mind, the purpose you just spent time pinning down.

If you discover that what your users want is completely different from what you initially had in mind, and you want to make that different thing, it is known as a product pivot. You can pivot your app if you want, but you have to throw out all the work you did on the old idea and start brand new in order to make a quality app. Don't hang on to old work just for the sake of having done the work. This is not like an hourly job; each hour you spend is working toward something. You're not counting the hours—you're counting the distance toward your goal.

If you keep your app's core purpose in mind, sometimes it's appropriate to ignore user feedback. This is counterintuitive, but it's also a principle that Apple Inc. lives and dies by.

> "A lot of times, people don't know what they want until you show it to them." Steve Jobs, *BusinessWeek*, May 25, 1998

This is an important point for you to keep in mind as well—the users know what problem they're having in their lives, but they're not always the most deft at identifying the appropriate solution.

If merely doing exactly what users say resulted in a good product, Apple would not be as successful as they are. Apple Inc. regularly ignores feature requests and has a limited scope of compatibility, which (next to price) is the most frequent complaint made about Apple's products. Yet, they still manage to be beloved and exclusively bought by millions of people, raking in billions of dollars.

So what feedback *does* Apple incorporate into their products? Apple looks at what the users want, not from a technical standpoint, but from a more emotional one. Apple asks what features would make the users feel good, not what features will make their lives better.

This means you should ask the users how they currently feel, not what features they want. Often the best features have never crossed the users' minds, but then revolutionize their lives.

This is evident on Apple's website. Go to the product page of any Apple product, and it won't have a laundry list of features and comparisons to competitors. It will have beautiful pictures of the product and emotion-evoking statements about it. Nothing more.

Tip Focus on what the users say the problem is, but don't necessarily implement the solutions they suggest.

Give the problem honest, concerted thought and come up with a solution (a core purpose for your app, as we discussed earlier) that evokes buzzwords like "innovative" and "intuitive". Find out how much your ideas or current solutions are failing, but don't get caught in the trap of obsessing over features.

The Final Test

The final test for app idea development is when someone will inevitably ask about your app. You must be able to answer these questions clearly, in under two sentences:

- What is your app?
- What does it do?
- Why should I download it over competitors' apps?

Until you can answer these questions, your idea isn't good enough.

Keep refining your app idea and keep soliciting feedback, until you can always clearly answer these questions in a way that fully satisfies the person asking.

When you tell your app idea to someone, they shouldn't be able to ask "Why wouldn't I just do this?" or "Wait, how does it work?" afterward. They should understand the first time you explain it.

If you do get the same follow-up questions over and over, hit the drawing board again. People should be able to figure out the answers without your help. Also, you don't want an app you continually have to explain to people over and over.

Don't forget that all of this feedback is so you can get people to pay for your app. So while feedback like "This is a great idea" is gratifying, it is not the same as a commitment to buy. If someone says, "I would buy this," write down their contact info. Their feedback is a lot more valuable to you than feedback from someone who is not a potential user.

In the next chapter, we take a look at how to make a website for your app.

CHAPTER 2

Website

One of the first things you're going to need is a website for your app—you need this before you've even made your app. Websites are free and easy to make. All you need is an app idea. It is not nearly as difficult as making an app.

Having a website isn't even technically necessary. Your app will already have a built-in website—your iTunes URL—and you could just provide that URL whenever people ask to view your app online. But making a website for your app is free and easy and confers a lot of benefits:

- Gives you something to promote while your app is in development
- Gives you a place to build a newsletter and audience
- Gives your social media somewhere to link
- Gives people something to share on social networks

The right time to build a website for your app is *right now*. Even if your app is nothing more than an idea, having a landing page and an email collection field allows you to begin building an audience as quickly as possible.

If you're not sure you even *want* to make your app, building a quick and free website can help you gauge whether you want to move forward with the idea. It doesn't cost any money and can help you determine whether your app is a good idea early on.

M. Holstein, *iPhone App Design for Entrepreneurs*,
https://doi.org/10.1007/978-1-4842-4285-8_2

Don't worry about making your website look perfect; the look of your website is going to change as time goes on, and users won't notice a change in your website. Your goal right now is to get something functional up for free, not to build the world's most polished sales website.

What Your Website Should Have

Before you can make an informed choice on how to build your website, you need to know what your website should include. It should contain everything someone would need to know to make the decision to download (or publicize) your app. Functions like support and company information should be far less obvious.

Note Your website's goal is primarily to be a big online poster for your app.

All of the information you need to display about your app can fit on one page, so don't hide anything behind extra windows or tabs. Don't make users go on duck hunts to find the information they need.

Your website must have the following:

- A big, obvious "Buy on the App Store" or "Available Soon on the App Store" button (see Figure 2-1). Apple has already spent tons of time and money building a visual association between an app and this button for you. Take advantage of this symbolic groundwork they already laid.

Figure 2-1. *The available on the app store button provided by Apple*

- An email collection field, to collect people's emails for your newsletter. Mailing lists serve tons of purposes throughout the app development process, and it's not an asset you can build overnight. You can add an email collection field easily with Sumo.
- If you have any reviews or comments about your app yet, include those, with a visual indicating where they are.
- If you have screenshots, whether they are mockups or real screenshots, include them. Users can't tell the difference between a real screenshot and a mockup, and they won't care. Make sure they're large, high definition, and everywhere. If you don't have screenshots, don't worry about it.
- A support link or customer support portal, so customers can contact you to resolve problems. This is the only section that should be hidden behind a link or button, although it should be obvious to users where the customer support portal is.

A fancy, professional URL was intentionally left off of this list. You could pay the $8 or $10 it takes to get a specialized URL, but you are at such an early stage in the development process that you may not know for sure what your app name or feel is going to be, and you don't want to spend money on a URL just for it to be useless two months from now.

Additionally, it's a non-essential expense, something you want to avoid until you're sure your app will sell. That way, if it doesn't sell, you haven't lost a ton of money.

Building Your Website

There are three services we recommend you use to build your website, and each has a different amount of capabilities.

LaunchRock (FREE)

LaunchRock is a service for creating free websites. LaunchRock is an interesting service because it's a platform like Kickstarter. You create a landing page on the LaunchRock website and record your progress creating the app and share it with the Internet. Share this page with potential users, and they can get an overview of the app and feel like they're in on the ground floor of your app's development.

LaunchRock is not specifically for apps, but they have templates specifically for app design. Make an account, select a template, upload screenshots of your app, and you have a website. LaunchRock also allows you to collect emails, like Squarespace, a necessary benefit to build a pre-sales list.

LaunchRock also has built-in analytics for the website, calculating total shares, hits to the website, and email list signups. Additionally, their templates are made for viewing on mobile and other devices, meaning you don't have to make your website mobile.

Squarespace ($99/yr)

Squarespace has been getting a lot of attention lately, as it has been a leading up-and-coming web hosting platform. They have quality website controls. It is exceedingly easy to do complicated things, and it is easy to maintain your website as well. In addition, they provide classy effects that give the impression of a website that was much more expensive to build than it really was.

Squarespace, being a full-fledged website builder, has a lot more power and flexibility than LaunchRock. If you want your website to be more than an online poster for your app, take a look at Squarespace's hosting capability. Squarespace also has built-in mobile-ready templates and a whole host of capabilities (downloads, forms, analytics, and other modules).

WordPress (Prices Vary)

If you need a big and powerful website, WordPress is what you want to use. It has a library of website plugins available for your website, and it is a beloved tool for many website designers. Prices vary between the hosting services, but if you're reasonably technical and want your app's website to do something unique, this is the service for you.

No matter what you select to build your website, you need to hook your website up to Google Analytics. They provide statistics and tracking information about your website, so you can analyze the kind of people coming to your website and what they're doing while they're there. This is information that in the future can help you develop and market your app effectively. Check out Google's analytics training to get familiar with their system.

Customer Support

Quality customer support is often the difference between a lot of positive reviews and a lot of negative reviews. Users with problems that were resolved quickly are more likely to review your company positively, stating that you were willing to do what it took to make them happy. On the other hand, an unhappy user will rant and vent their frustration with a one star review. You need to get an efficient support system in place in order to secure those positive reviews.

Note Irate users with problems, when they get great service, can turn into the happiest and most satisfied users.

After you work with an unsatisfied customer over customer support, make sure to ask them to leave a review of your app as the final step of the support process (if the interaction was positive).

Your website (and an email link from iTunes) is how your users access your customer support. You need to have a customer support system set up as early as possible, so that you can effectively handle your customer support requests. You don't want to miss any tickets, or be inundated with them.

The easiest route might seem to pop your personal email on the website and iTunes listing, but this is not what you want to do. If your app hits any measure of success, your personal email will be inundated with complaining users' emails, filling up your inbox. Imagine trying to go about your regular life when your email is clogged with complaints, and you aren't able to filter out the emails that are actually important.

The better route is to create a free email account with Gmail, such as yourappname@gmail.com, but this isn't the best approach. While it separates your email accounts (good), it doesn't have any tools to organize or handle mass amounts of emails (bad).

There is an ideal and free solution to these problems. There are websites and services out there that create whole solutions around handling customer support emails. These customer support portals will track data, links, individual customers, the time it took you to respond, and every other imaginable metric. That way, you can stay on top of what's happening with customer support.

These also plug in with free website builders, since all you have to do is include a link to your customer support portal at the bottom of your website under "contact us" or "customer support". No technical expertise is required to install them. The best part is that every one of these services has a free plan level.

- *UserVoice*: If you want a free customer support portal, UserVoice is your best bet. For free, you can get an entire ticketing and email handling system, intricate metrics on who contacts your customer support, and a unique but powerful forum setup that allows users to vote on ideas. This is fantastic for finding out what features your users want you to include, which can be difficult information to get. Additionally, they have a full knowledgebase plugin, so you can post how-tos and support right on their website.
- *Freshdesk*: Another favorite, Freshdesk scales much more quickly and easily than UserVoice does. If you're willing to shell out the cash, they have more customization options, so that you can make your Freshdesk portal look like your app's website. They also have more features, if you have something special in mind for your customers. At the free price point, they have fewer features than UserVoice, but at the paid levels, they have more.
- *Zoho*: Zoho support has domain mapping at the free price point, which means your Zoho customer support portal can have the same URL as your website. This is not a common feature at free price points. Zoho makes a range of other apps as well, which means that you have the added benefit of products that integrate entirely with each other, saving you time.

In the next chapter, we take a look at how to find out if your app will sell on the App Store before you've even built it.

CHAPTER 3

Idea Validation

You need to make sure people will buy your app before building it, because you liking an idea doesn't mean lots of people will like an idea. Even if a few others have told you the idea is brilliant, that doesn't mean it will actually sell well.

Take, for instance, celery. Most people don't like celery, but celery-lovers do exist. If your app idea is celery, that's fine. Not everyone is making an app for the masses. Some apps, like bird song libraries and Spanish-Portuguese dictionaries, will just not be needed by everyone. What *isn't* fine is assuming your app is cake, when your app may very well be celery. How can you make sure your app isn't celery?

Note We touched on getting feedback from people in the last chapter, but that was designed to help you hone your idea. Real life conversations with people who are right here in front of you are great for developing an idea, but not so much for making sure it will be popular.

You can find out whether or not your app will be successful by putting it online and seeing if it actually is. In other words, you need to go presell your app. This is very simple; post a link to your website on forums where your intended users are active. See how well your website is received, how many email signups you get, and what their feedback is like.

M. Holstein, *iPhone App Design for Entrepreneurs*,
https://doi.org/10.1007/978-1-4842-4285-8_3

Promote Your Website

So, you need to post a link to your website where your intended users are active. Here are some places you can look for those users:

- Reddit is a forum website with a category for everything under the sun. Different categories are called subreddits, and you can search through them on Reddit. Post a link to your website on all the related categories you can find.
- Hacker News is a great place to post if your app idea has wide appeal, such as a photo-editing app or a social network. It is essentially a tech news website, and high-tech apps will get traffic here.
- Use JustUnfollow to find leaders in your industry and follow them and tweet at them. Also follow anyone following accounts like yours, to grow your Twitter audience initially. This creates interaction on your Twitter page and gets you involved in the online community.
- Comment on articles that your target market is reading. People who are very interested go through comments on small and medium websites, and your website would make sense there.

When you post about your app, you don't want to be spammy. Spammers do not think they're spamming, and learning to tell when you personally are coming off as spammy is necessary. Here are some tips on how to avoid coming off as spam.

Tell a Story

It won't take you too long to come across a post that says "Hey, I'm creating this app called flim flam, will you download and review? Thanks!" People will say "nope" and scroll on by. They do this because there's no emotion or story attached to this post to reel them in.

To avoid people scrolling on by your post, make sure it says:

- Why the app idea excites you
- What your dreams are for this app
- How you had this idea
- Who you're trying to help with it

Give Readers a Call to Action

You've got readers reeled in and feeling emotions, and now you have to do something with it. What do you want your excited readers to do? Some things they can do are:

- Sign up for your mailing list
- Share comments or critiques of the idea
- Share some sage wisdom with you, if they have any

Keep It Personal

Don't be official or come off as marketing copy. At this point, you're a person on a journey. Make sure they're along for that journey by sharing your emotions and experiences with them. Ask viewers for advice on anything you're struggling with, or perhaps their top tip to help you out.

Interact with Signups

After you've promoted your website and built a mailing list, you need to interact with people who have signed up. There are a couple of ways to do this, discussed next.

Start a Company Blog

This can be easily done using a service like WordPress or Tumblr. Don't worry yet about connecting URLs or making your website and blog look nice. Right now, your worry is building an audience and proving that spending time and money on a website is worth the effort.

Writing blog posts provides a context for emailing your audience with information relevant to your app, and something for your audience to share with other people, thus increasing your presence on the Internet. Once you've written useful blog articles, share them with your audience. Send them out in newsletter updates and post them on your social media.

Tip To learn more about what a blog can do for you and how you can leverage that blog to increase subscribers and sales, check out Quick Sprout, OkDork, and Videofruit.

Ask Subscribers Questions

At the bottom of the emails you send to your signups, ask them questions. Ask them what they are most excited about in this area, what they'd love to see in an app like yours, or anything you want to know. Most people won't answer, but those who do tend to be willing to discuss their answers with you in more depth.

Survey Subscribers

You can create surveys for free and distribute them easily with SurveyMonkey and Google Docs. Surveys allow you to collect information about what your users want from your app. This ensures that the product you build will be something that people love, not just something you hope they love.

How to Create Your Own Survey

Start your survey by asking about demographics—age, location, gender, occupation, and so on. Knowing your market is critical, and sometimes your primary demographic doesn't turn out to be what you think it is.

Make questions multiple-choice wherever possible. People don't like writing out answers to things, and if they're multiple choice you can get statistics on the results as well. Always provide an "other" option and a free-response textbox, so they can add comments if they have any. You don't want to miss out on feedback because of a forgotten textbox.

Use simple language and short sentences. If respondents need to think more than a half-second, they will stop and think it's not worth their time. Smooth over the process for them as much as possible. In the same vein, a shorter survey is always advised as well.

Ask respondents how good their current solution for this problem is. Whatever app you're making, someone somewhere has also tried to solve this problem, and asking how well the responder likes their current solution tells you how the current solution is doing.

Five stars shouldn't merely be "it did everything I wanted". Five stars is a product they should be stark raving mad about, and that is your goal for how they'll feel about your app. Make this clear to respondents. If your user is satisfied with their current solution, it should only be a four-star rating; that way you will know if your users are stark raving mad about their current solution. Three or fewer stars indicates a market gap that you could easily enter.

Ask Users Good Questions

The bulk of this section is inspired by a *Medium* article that perfectly addressed this problem. Ask your users good questions. The goal of interacting with all of these email signups is to get them excited about your app and to get their feedback. Gathering feedback is the best way to get direction for design, because you are designing for the users. Directly interviewing these users has a couple of key advantages:

- It's free, which is always a huge advantage
- It's quick (takes talking to five or fewer people)
- You often get more than you ask for

However, it's easy to get it wrong. You can't just go up and ask, "would you like an app that does x"—that question yields a lot of false positives, because it's a yes/no question where someone's feelings may get hurt.

Visionaries like Steve Jobs know that people don't know what they want, because when you ask someone what they want, you get what they envision as their ideal solution—not the actual ideal solution. This means if you go right up to people and ask "What product do you think would solve your problem," the answers are going to be next to useless.

Note Think of the traditional buggy vs. car example. If Henry Ford asked his customers what they wanted, they would have said "a better horse". But we might all agree that we're happier that Ford made the car instead.

The Questions You Should be Asking

This section discusses the questions to ask your users.

What are You Trying to Get Done? (Gather Context)

Most of the time, someone's initial answer to this question isn't going to be the answer that you're looking for. It will appear at first glance that it is, but sometimes what people *think* they want is based on a bunch of faulty assumptions that they want. In order to get through all of this confusion, make sure to ask why several times to get to the root of the matter.

For an example, take a look at the user interaction outlined in Figure 3-1.

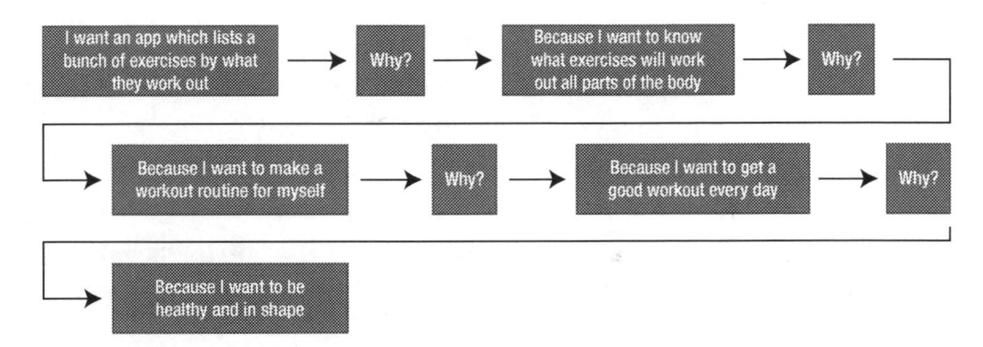

Figure 3-1. *An example user reasoning out their real need*

The user in this case thought her problem was that an app categorizing exercises by body part didn't exist, but her real goal was to be fit and healthy. There's probably a simpler way for her to accomplish this goal than the process outlined here. Now, instead of creating an app that catalogues exercise routines, you can create an app that creates a workout routine based on the user's current body and goals, without mucking about with the whole process. That would make this user much happier, and isn't what she thought she wanted?

Tip You want a "something" that you can build an app around, and usually this isn't the first answer someone provides.

This can also reveal user problems that are due to a lack of understanding or incorrect use, as opposed to a solution not actually existing. Then, if you wanted to solve the problem, it would be by educating people on the solutions that actually exist or creating one that's easier to understand.

How Do You Currently Do This? (Analyze Workflow)

To solve users problems better than they currently do, you must first understand how they solve their own problems. Knowing the full extent of the user workflow puts you in their shoes and gives you the information you need to do a better job solving their problems.

Understanding their workflow allows you to understand how to simplify their solution. Even though they might not necessarily have a problem, there will likely be parts of their workflow that are inefficient and unnecessary. You should minimize those. Then, you can focus on making the most important parts of the workflow better.

What Could be Better About How You Do This? (Find Opportunities)

After you've asked the first two questions, you should have a pretty good idea of what is going on with your users and should have a hypothesis for how you can fix their problem. When you ask this question, you can verify your hypothesis. If you were warm, proceed—if you were cold, go back to start and try again.

In the next chapter, we take a look at how your app is going to make money.

CHAPTER 4

Business Models

Once you've ironed out what your app is actually going to be, you need to decide how you're going to make money off of it. In other words, you need to decide how your app company is going to work. This is the essence of creating a business model.

Defining your business model means thinking through how you're going to organize your app, how you're going to make money, who your target users are, etc. Because really, making an app is launching a (very) small, part-time business.

Making a business plan is not the complicated, 10-100 page endeavor it used to be. With the upsurge in entrepreneurship in American culture, tons of easy business-modeling tools have become available.

The tool recommended to you here is made by BMFiddle. They provide a tool called the *business model canvas,* which is a visual model for making a business plan. This makes creating a business model as easy as filling in the blanks. BMFiddle has a resource section that you can explore to get up and running quickly.

In order to create a good business model, you should understand the two different "typical" business models for app businesses. Chances are your app idea falls into one of these common business models.

M. Holstein, *iPhone App Design for Entrepreneurs,*
https://doi.org/10.1007/978-1-4842-4285-8_4

The Long Tail

The *long tail* model is a business model where one producer makes many apps and promotes their apps inside their other apps (known as *cross-promotion*). Since there are many apps, each app represents an opportunity to get more money from one person. For this to work, all the apps must appeal to the same audience base you're building.

It is predicated on duplicating what become popular over and over, and counting on people to buy these copycat or off-brand apps over and over. Each popular app duplicate will be all very similar (or completely the same), with different graphics applied to the app to make it look as if it's a totally different app.

It is looked down upon by some app developers as not being very dedicated to what you do,—and more about making a quick buck—since it is these kinds of apps that lead to App Store overpopulation. Sure, there's technically more money to be made and you're technically adding sales, but you're not really adding anything more to the App Store.

This is the model that made Chad Muerta, a well-known app developer, wealthy from app development. Other examples of this include the multitude of app copies of Flappy Bird, once Flappy Bird became popular on the App Store, or the multitude of app copies of Angry Birds.

Pros

- There is a higher amount of money to be made from each user quickly, since you can push out each new app easily.
- Each app does not need to be 100% amazing, polished, and perfect.

- The chances of you making more money go up. The median app doesn't make back $500, so the more apps you make, the better your chances are to beat those numbers.
- You only have to do the hard work of coding the actual app once, and you can resubmit it over and over.

Cons

- No best-selling apps come from the long tail method, due to the fact that you'll have so many apps you need to focus on. Quality work comes from focus.
- Copycat apps may hit the top charts temporarily, but except for the Flappy Bird fiasco in which the original was removed from the store, the copycats don't stay high-ranking very long, and are looked down upon.
- Upkeep on many apps is enormous, and you'll almost definitely get behind on bug fixes. You don't want to be the face behind poor and out-of-date products.
- As you make one app better and better, you'll get enthusiastic users who will certainly be willing to pay for iPad versions and expansions, or who will accept a price increase when you've increased the value of your software. You'll never get users with the same enthusiasm in the long tail method.

The Flagship

This model is exactly the opposite of the long tail model. You produce one app and pour all of your effort into this single app. It is the one beautiful app that receives great design, great code, and great marketing. It is the type of business model that makes millionaires.

It's a respected business model, by and large, and is the one that tech startups use when launching a serious venture. If you are looking to make an entrance into startup culture, this is your best bet.

A business model can have multiple apps and still technically be in this category if each app has its own independent codebase, website, and marketing materials.

Pros

- This model provides a real chance at producing a hit, million-dollar app that tops the charts and stays there. It's no guarantee, but unlike the lottery, you have a better chance than the next guy.
- You only have to keep one app up to date, so you'll never get behind on progress as technology marches forward. Your app can more easily stay cutting edge.
- As you make one app better and better, you'll get enthusiastic users who will certainly be willing to pay for iPad versions and expansions, or who will accept a price increase when you've increased the value of your software.
- A good product always gets downloaded more.

Cons

- In the beginning, when your app is not big and beefy with features yet, there's less money to be made from each individual purchaser.
- If you only have one app, you can't give it half your effort. You need to commit fully.
- If this app ends up being a lackluster market dud, you need to beat away at this app's core offering over and over until you get it right.
- Each update requires real new code—no copying or reusing code from your other apps.

To decide which business model is going to perform the best, you need to consider who your customer is. Not just a broad, general statement, but a specific picture of your model user.

Model User

A model user is a picture of the hypothetical average user of your app. You create a model user so that you can then make assumptions based on this model user, such as where you need to advertise, what you need to change, what the price of your app should be, and how many apps you should make. This model user is the exact person you began designing this app for.

If you try to make an app suitable for a large category, you'll end being too general and nobody will be interested in your product.

For example, say you're looking for an app that acts as a level, so that you can hang a shelf up in your house correctly. If you find an app that says "for home construction projects of any level!" and an app that says "helps you hang things level in your home," you're going to download the second one because you know it does what you need. Thus, the developer with the more specific user in mind ended up getting the download.

Once you've completely dominated the market of your original target user, you can move on to bigger and better markets—this is how most recent global companies began. They picked a niche market and became the market leader, and then they used that momentum to move on to bigger markets. It's a cliché, that "niche makes you rich".

Facebook started off as a social network website only for college students, but then after they dominated the college student market, they moved onto young adults and high school students. After they dominated those markets, they set their sights on everyone at large.

When you're identifying your model user, here are the things you should know specifically about them:

- Age
- Gender
- Favorite websites (for news related to your app's industry)
- Favorite websites (overall)
- Clothes (what are they trying to communicate about themselves?)

Your model user will determine which basic business model is best for you. First, for each individual app, you need to determine your target user for that app. If all of the target users are similar or exactly the same, you should employ the long tail business model. If you have many different target users or you have only one app, the flagship model is best for you.

Every business model is going to be one of the two, or a mix. Nail down what your business model is now, so you can make choices about company names and marketing ahead of time. To make this decision, think about who your customer is and what they need the most.

In the next chapter, we're going to take a look at how much you should charge for your app.

CHAPTER 5

Fair Pricing

After you've decided on a business model, you need to determine how you'll make money. Picking a good pricing model for your app is difficult. The App Store is a highly volatile market, and the best pricing model can change as quickly as the App Store itself. Additionally, price wars have driven the median price of an app down to $0.99, so it can be hard to justify an app that's more expensive than that.

Don't get caught up in the price wars of the App Store. You will get fewer downloads at a higher price and some people may scoff, but the extra money from a higher price will more than make up for it. Most of all, pick the price that you want. If you don't feel comfortable charging less than $5 for your app, don't.

Tip Give your app the price it deserves.

As of right now on the App Store, there are three methods of pricing your app—You can make it a flat cost to download, you can make it *freemium*, which is free with a premium version, or you can have an in-app purchase option to unlock the extended features. Let's discuss the advantages or disadvantages of each method, as well as the techniques for implementing each.

M. Holstein, *iPhone App Design for Entrepreneurs*,
https://doi.org/10.1007/978-1-4842-4285-8_5

Flat Pricing

Flat pricing is setting a download price for your app, and once the price is paid, the user can download your app. This is the only pricing model in which no additional code has to be written to use it. When you have to pay money to download an app, it is flat pricing.

Implementing this pricing comes with two different options for price strategy, depending on how you want to vary the price.

High-to-Low Pricing

One strategy to go is to start your price high, and then drop the price lower and lower as you penetrate the market and people begin to recognize your app.

Pros

- The price drops as your app becomes more outdated, and you can raise the price again upon subsequent large updates.
- You can lower the price to discount your app. For instance, a $1.99 app can have a Halloween or Christmas special, whereas a $0.99 app can only be discounted by giving it away. There are a great many people who believe you simply shouldn't have a free or $0.99 app because of this advantage.

Cons

- Your app is already more expensive than a lot of the herd when it's released, making your entry into the App Store less impactful.

- An app can never explode into success if it doesn't gain the critical amount of users to get the word-of-mouth marketing rolling. Gaining enough users for this will be harder with a more expensive app.

Low-to-High Pricing

You can also start your app at a lower price, and then increase it as your app lives on.

Pros

- This is good for apps that are introduced as very small and simple apps, but as features are added and the scope is expanded, they end up providing more value. You wouldn't pay the same amount for a simple individual's to-do app as for a project management app that allows global real-time team collaboration across devices. Developers should be conscious of this.
- Staring low can get you the exposure you need right off the bat, so your app doesn't languish because it was never seen in the first place.

Cons

- It's going to take you longer to earn back your initial investment. Be prepared for more financial struggle.
- There's no space to discount your app. You can never run a discount promotion.

There is also an interesting relationship between $1.99 and $0.99 pricing. Anecdotal stories from app developers show that revenue is equal for both price points, but when an app is priced at $1.99, its reviews are better.

Additionally, the kind of people willing to pay for apps aren't going to blink at the difference between one and two dollars. Take this with a grain of salt as no official studies have been done, but it makes sense.

One theory is that people who spend $1.99 perceive themselves as having spent more than $0.99, so they want to justify their spending. They don't want to have bought a bad app for that amount of money, so even if it is poorer than they expect, they aren't as critical. They look for the higher value that they paid for. However, $0.99 is cheap, less than a dollar, so if they did get a dud, they are happy to complain about it. These people are also less likely to pay for apps in general, which is why they stick to $0.99 or lower price points.

Know that there are no general rules about which price points do and don't work. Whether or not a pricing model works depends highly on your situation. You'll need to investigate your competitors and compare your product to theirs to find the right price. Figure out where you stand next to your competitors and price your app accordingly. You'll also need to think about your model users, and what price they'd be willing to pay. If they're wealthy or are buying your app for work, they're willing to pay more money than if they are a teenager looking for a time-wasting game.

Additionally, the right price for you will change over the course of a year and over several years. Don't be afraid to experiment with several different prices in order to see which ones end up making you the most money. Users will not notice price changes on the App Store unless they hit double digits, or you run a very large company. Experiment by varying the price over time until you find a price point that makes you the most profit.

Freemium

Freemium pricing is when your app is offered for free, but you have in-app purchases available. Freemium pricing is the most popular pricing structure on the Games section of the App Store. And it's no wonder, as games are perfectly suited for it. You offer the first couple of game levels for free, to get them hooked on the game. After users are hooked, they're much more willing to pay a dollar or so for extra levels. This is proven on the App Store to be much more effective than charging a flat price for a game.

Some games also offer all levels for free, but you can pay a certain amount of money to skip the level. Or, games offer in-game bonuses for purchase (bombs, superpowers, weapons, etc.). This content is consumable, meaning people can keep buying this content over and over, and you can make a virtually unlimited amount of money from each individual player. Most people will not buy any of this content, but if 1%-5% of your users spend upwards of $20 on a single game, that's a lot of income per person on the App Store.

The freemium model isn't just used with games. Many apps are moving to this pricing structure, with extended features available in many other free apps for a subscription in-app purchase. This is marketed as the "pro," "full," or "extended" versions.

This pricing structure has the benefit of initially being free, so many people click on the button and download the app, getting it into a lot of people's hands and trying it out. If you have an excellent app, this will increase the amount of revenue you make, since your users have confirmed it is useful and are more willing to spend their money.

However, if your app isn't top-of-the-heap polished and stunning, the freemium model can hurt you more than it can help. Word will get out about the quality of your app quickly, with so many people downloading it and reviewing it. The jig will be up, and people won't pay for in-app purchases. That said, failing quickly is always much better than failing slowly, as it allows you to incorporate what you've learned and try again without wasting time. Hence, the entrepreneur cliché—"fail fast, fail often".

If your app relies heavily on user-generated content or social media such as Instagram, Facebook, or Twitter, a free model is almost certainly the way to go. That social media and user content has to be created by the masses using your app, and typical conversion rates for in-app purchases are only between 1% and 5%. Therefore, you need more people than that using your app if you expect to get the amount of user-generated content you need.

If your app is a personal, non-addictive experience, you may be better off using a flat pricing model. If people already have access to limited usable functionality and don't find themselves with a burning desire for additional functionality, they may find going without it is worth the money. And if they have to pay outright (instead of an in-app purchases), the experience of the app is more cohesive and whole once it's downloaded, which can be worth a lot to people with very busy or full lives. Consider flat pricing for the Business, Productivity, Navigation, Finance, and similar app categories.

In-App Advertisements

The third option is using in-app advertisements. iOS apps can have subtle and non-invasive in-app advertisements, and these can help monetize an otherwise small app, such as a utility or calculator app. They can also help monetize the free version of addicting apps, as seen in many games produced by Zynga.

Advertisements can be a very good option for an app that targets a demographic with either less spending money (low-income households or young college students) or no money (high school, middle school, and elementary school students). While people in these demographics cannot give you money, they are happy to give you their and their friends' eyes, which advertisers will pay for.

For this financial model to earn significant income, you need a lot of eyes on your app—Flappy Bird made $50,000 a day because it got millions and millions of eyeballs on it each day, which only the top 1% of apps can do. That said, only a free app can do that—no paid app will ever get that many eyeballs on it. Only free apps can become such big hits.

As long as you integrate the advertisements in a graceful way, they can make a good app available to everyone. However, advertisements need to be considered in the design of the app itself, from step one. They need to be integrated well, or people will throw your app away in frustration.

Picking a Pricing Model

Your goals for your app will determine your pricing decision, along with what your competitors are doing and what your users expect. Do you want to be a well known developer with a famous app (but maybe not as much money), or are you content to languish in obscurity if you're supporting yourself with your app's earnings? Your pricing model should reflect your personal goals. You can support one expensive app with fewer users, or support a less expensive app with hundreds of thousands of users.

Your pricing should also reflect your position on the App Store. Niche markets have pricier apps because there are so few people buying them. Because demand is low, there aren't as many suppliers, and because there are not as many suppliers, price goes up. Charging any less would cause developers like you to lose money, and the apps to get discontinued. However, mass-download markets such as games have $0.99 or free apps because many, many people are downloading them, and a high price would drive users away to other games instead. The larger your market, the lower your price should be.

Whatever you decide to do, you need to remember to always vary your price based on what the market says. In order for your app to perform the best, you should vary the price throughout your app's lifetime. Here are some guidelines for price adjustment:

- If you're using a flat pricing model and enjoying a large and consistent amount of downloads, up your price to increase your revenue. If this decreases your total revenue, lower your price below your original starting point, and that should bring in more revenue.
- If too few people are buying your app, lower your price and let your app penetrate the market before raising the price again. If that only decreases revenue further, increase your app's price to find the appropriate equilibrium.

Your app shouldn't stay the same price during its lifetime. The reason pricing an app is such a difficult art is because the optimal price for any given app is constantly in flux, which means you must constantly be changing your app's price.

You build your apps for the people you're helping, but you have to sustain yourself as well. You can't build fantastic apps for people if you're busy working your day jobs and coding for absolutely free in the night, losing sleep and deteriorating your own health. That's a great short-term strategy to validate an idea, but your app would have to be discontinued if you kept on this way. Everyone dreams of quitting their day job and being the next app millionaire, but without fair pricing that can't happen.

> Matt Gemmell, of Instinctive Code:
>
> @mattgemmell: Making more money and getting rid of people who don't think your app is worth much: is there anything that fair pricing CAN'T do?

In the next chapter, we'll start designing your app.

CHAPTER 6

Planning Your App

Hopefully the last section in this book gave you a solid idea of what you want your app to be. Not only do you have a core mission for your app that is aligned with what users want, you have identified the problems they experience in their lives.

This means that you've defined all of the benefits that the users need, but not the features that those translate into. *Benefits* are what they need to achieve; *features* are how you get there. In tech jargon, this is known as a use case. Provide examples of a use case.

As you translate the idea and use cases you've defined into your final feature specs, keep the golden rule of app design at the front of your mind: *Don't have too many features.* This point is being emphasized again because it applies here too. You already know how too many features can be bad for your users and your app idea, but it can also be the end of your project.

Some pitfalls you should take care to avoid:

- One of the biggest pitfalls new entrepreneurs fall into is the need to make their product perfect before releasing it. You don't want to release an app with bugs, but you don't want to be too slow to market because you're trying to make the first version too many things. This means not packing too many features into your first version. No famous app companies today have apps that look anything like their version 1. For example, Figure 6-1 shows Facebook then and now.

M. Holstein, *iPhone App Design for Entrepreneurs*,
https://doi.org/10.1007/978-1-4842-4285-8_6

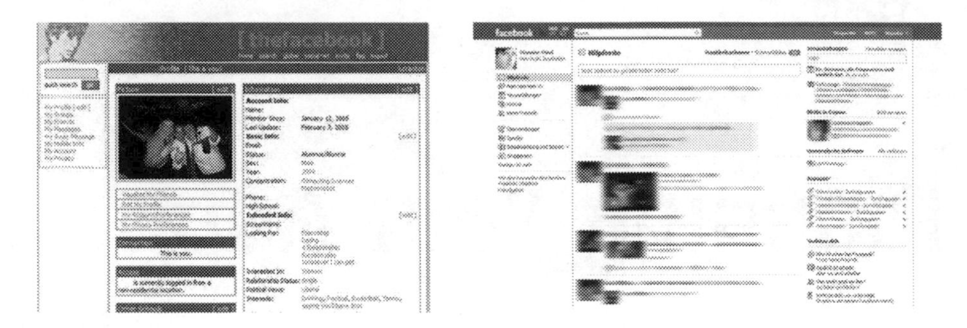

***Figure 6-1.** Facebook's first version versus Facebook in 2016*

- Another danger is that if you add features before you have a user base, you may find you're adding features your users don't want. Nothing sucks more than slaving away at a tough or complicated feature, only to find out your users aren't using it. If you're making this app to improve their lives (which you are), you will want to solicit their input before adding features.

All of these guidelines are to help you make what is called a "minimum viable product," and it is what it sounds like—the smallest, easiest product you can make. Release it, get feedback from users and assess the situation, and add features with each release. That's the core of the MVP philosophy, (popularized by Eric Ries in *The Lean Startup*), along with validating your idea as you did in earlier chapters. This means that you won't stray too far from what your users want, you will be quick to market, and you won't lose your own personal momentum.

App Outline

Having an outline can save you a ton of time and frustration designing your app and help you nail down which features you want. It is a plan for making your app's layout. It is astonishingly easy to get done with all your wireframes and all your designs, only to find you left a Delete button out and have to start from scratch to make it work. This is the kind of mistake you may make if you begin your wireframes now.

You'll know you're done with the outline when you can go no deeper and can break down the basic functions of the app no further. The purpose this serves you is that of organization; you quickly identify features that you need —such as deleting or saving—and the outline helps organize every technical part of the app that needs addressed. This level of detail is not for sharing with the users, but for sharing with yourself and your development team. With this outline, you know what features each screen needs, so when you wireframe your app you won't forget any of them.

Figure 6-2 shows a simplified example of an outline for an app called Visual Routine, which is designed to make routines for children.

- Make Routines
- ~Make, Delete, & Autosave A Step
- ~~Make, Delete, & Autosave Photo
- ~~Make, Delete, & Autosave Text
- ~~Make, Delete, & Autosave Audio
- ~~Make, Delete, & Autosave Choices
- ~~Make, Delete, & Autosave Photo
- ~~~Selecting a stock photo
- ~~~Selecting a photo from the camera roll
- ~~~Taking a photo with the camera
- ~~~Deleting the current photo
- ~~Make, Delete, & Autosave Text
- ~~Make, Delete, & Autosave Audio
- ~~~Recording new audio

- ~~~~Playing back audio to test
- Autosaving a Routine
- Use Routines
- ~Open Routine
- ~Play Audio
- ~Mark Cells Completed
- ~Mark Cells Incomplete
- Share Routines
- ~Export
- ~~As .PDF
- ~~As Images
- ~~As App Filetype

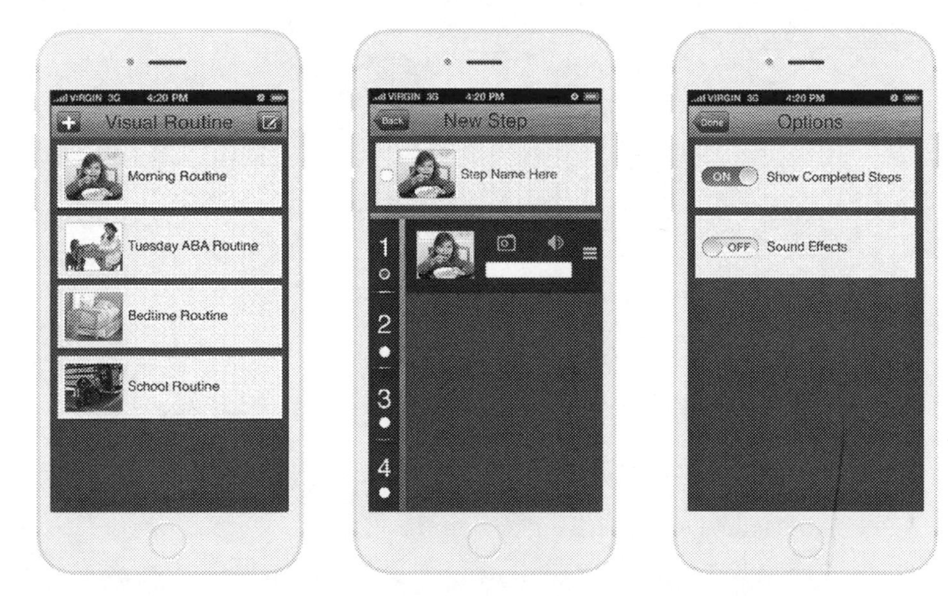

Figure 6-2. *Visual Routine app, 2016*

This outline also makes it easy to transition into writing pseudocode for your app. Much like you write an outline before writing a paper for school, you write pseudocode as a draft for your app. This draft is necessary for whoever programs it. Here's an example of pseudocode for the same app, Visual Routine:

- Tap a routine; routine opens
- ~if the photo is tapped
- ~~play the audio
- ~~~if there is none, do nothing.
- if the checkbox is tapped
- ~if it is complete, mark incomplete
- ~~if it is incomplete, mark complete
- ~~~if the audio is enabled, play 'step completed' chime

Pseudocode is a great tool no matter what type of development you're using. If you're coding your app yourself, you can use it to ask people for help when you get stuck, and you can use it to plan the most efficient code and algorithms for functions. If you're hiring someone else to do it for you, you can give it to them so they know exactly what to code and what you're looking for when they're working.

Gestures

"What did you just do?" "This?"

Four fingers swipe across the screen, and it switches to the next app.

"Yeah! That!"

People know astonishingly little about the basic gestures on iOS, even this many years after the iPhone's release. The average person on iOS knows only the most obvious—scrolling up and down with a movement of the fingertip, swiping sideways to turn a page, and if we're lucky, the slide right-to-left to delete.

If you're not convinced people don't know gestures, check out the Bing page covering "iPhone Gestures":

- Five Secret iOS Gestures You Need to Know
- Learn All the iPhone Gestures Here
- Top 5 Secret Gesture Tips

All of these articles link to what power users would consider fairly basic gestures. Figure 6-3 shows a chart of many different gestures, all of which can be found on different versions of iOS and iPhone apps.

Figure 6-3. *Common gestures for mobile devices*

If you don't know all these gestures, you can't possibly expect your users to know them either. This isn't to rag on people—there isn't anything inherently superior to knowing more gestures, and folks learn the gestures as they need them to successfully get through their lives. It is up to you, the designer, to empathize with your users and design a product for them, which includes gestures best for them.

Most people alive are too used to the mouse-computer metaphor to adopt fancy or complicated gestures, and those who aren't are barely entering high school. Unless your app is designed for those individuals, listen up.

Don't hide features of your app in a rotating or multi-touch gesture, because many people will miss them. If you include a fancy or three-finger gesture in your app, provide users with a way to do the action without having to do the gesture, or provide very obvious prompting the first few times. Being on the cutting edge can alienate people. You can include gestures, but make sure people see the physical metaphor of what they're doing. For example, to swipe, you need to be prompted by the sight of a page.

For more complicated gestures, if you don't explicitly tell users what to do upon opening the app *or* provide an alternate visually-based way of performing the action, it will simply be lost on them. A lot of apps take the hybrid route of including the more complicated gestures for the power user but still provide a seamless visual way to do something for the casual users.

Stick to Apple's Gesture Guidelines

People are already natively familiar with the pinch, tap, and rotating of the iPhone, if not from personal use, then from seeing commercials. For the few gestures like these that people do use frequently, you need to be careful not to rewrite what a gesture means in their head.

For instance, swiping down from the top of the screen brings down the notification center; creating a conflicting gesture will both confuse the iPhone on input and confuse the user.

A more complete list of gestures can be found in the *iOS Human Interface Guidelines* online, which will help you decide which gestures belong on your iPhone app.

Note If people can jump into your app and understand your unique and new gestures intuitively, you have design gold.

Using Animation

When used subtly, animation can make an app looked polished and professional. It can make a boring app look classy. But if overdone, it can ruin an app like a tacky Flash game. There are a couple of circumstances where an animation is appropriate.

Loading Screens

Thanks to the lack of reliability of computers in the past, a still screen is now associated with a crashed program. This means when something is loading, the screen must remain in motion to assure the user that the app has not crashed. You can communicate this with either a loading wheel or loading bar, and which one is better depends on the situation.

When it will take more than a couple seconds, bars are recommended over wheels because they indicate the extent of the progress. Like being put on hold when calling your insurance company, a wheel gives absolutely no indication of when it will be finished, which frustrates the user in and of itself.

When the loading time is less than a couple of seconds, a spinning element is recommended over a bar. The bar will track to the end so quickly that it's not worth the effort it takes to create one, and a spinning element will communicate the loading message much more quickly.

If you wanted to add interest in an especially long loading screen, it could provide fun facts related to the app, such as fun state facts for an educational app or productivity tips for a to-do app.

Real-World Metaphors

Animation is also appropriate when there is a real-world equivalent to the action you're trying to take. Like in the example of a book page turning, animation can be suitable when it references a real-life equivalent to the app.

An excellent example of this is Apple's iBooks app. There is a page-turning animation when reading a book in iBooks, but it is extraordinarily quick and easy to use. The speed of the animation matches the speed with which you swipe a page. This brought the app up several notches and made it seem palatable to an audience that was still suspicious of ebooks. See Figure 6-4.

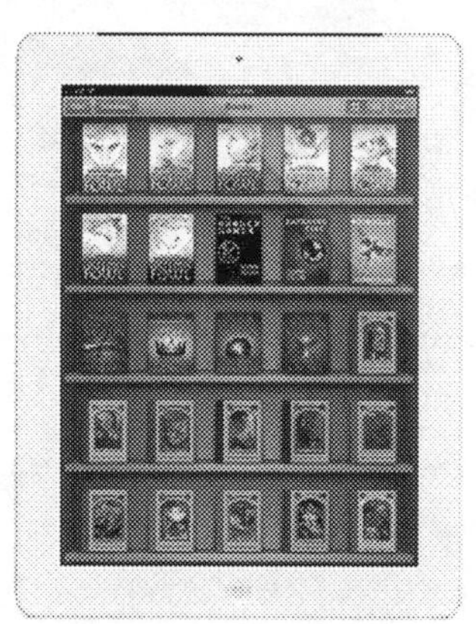

Figure 6-4. *Apple iBooks, 2016*

With real-world animations, you have to be careful not to take them too far. Imagine if Apple took these animations too far, making you watch a flourishing page every time you just wanted to flip back a page to reread something. This would get frustrating quickly, and you'd stop enjoying the ebook experience. Be careful not to overdo animation, as that can make your app tacky and difficult to use.

Screen Transitions

Occasionally animations can be appropriate as screen transitions, but they should be used sparingly. Much like with a PowerPoint presentation, tasteful transitions can make it that much better, but obnoxious transitions can ruin the slides.

There are some typical transitions that accompany different navigation styles throughout iOS. For instance, switching to the home screen or to different apps is a half-second fading transition, and navigating screens within an app using the header navigation is a sliding left or right. These transitions depend on the navigation you use.

For any design element, you can pursue Apple's far more thorough guidelines for animation in its *Human Interface Guidelines*.

CHAPTER 7

User Experience Design

The designing you're doing now and have been doing up until now is known as *User Experience Design*. It's a huge and growing field in software design, and it is why Apple's apps all work so smoothly. It's what you're going to learn to do so that you, too, can have a really great iOS app. So that you can talk fluently about what you're learning to do with other people, here's a bunch of basic jargon that is going to be used from here on out:

- *UX/User Experience*: UX is literally the experience the user has while using the app. Good user experience would be the users accomplishing what they intended to quickly and painlessly, and even enjoying themselves because of how simple it was. Bad user experience is getting lost within the app, not knowing how to do something, or tough controls to use. What UX is not is flat design, "responsive," or any other word for a visually appealing app. An app with good UX can be in black and white and still be a pleasure to use.

  A great example of this is an app called A Dark Room. It's a game with black and white graphics and plain text, and yet it was in the top charts of the

M. Holstein, *iPhone App Design for Entrepreneurs*,
https://doi.org/10.1007/978-1-4842-4285-8_7

App Store for quite some time. It remained in the top charts so long because it has a fantastic user experience. People keep going back to it, despite the web 2.0 black and white interface pictured in Figure 7-1.

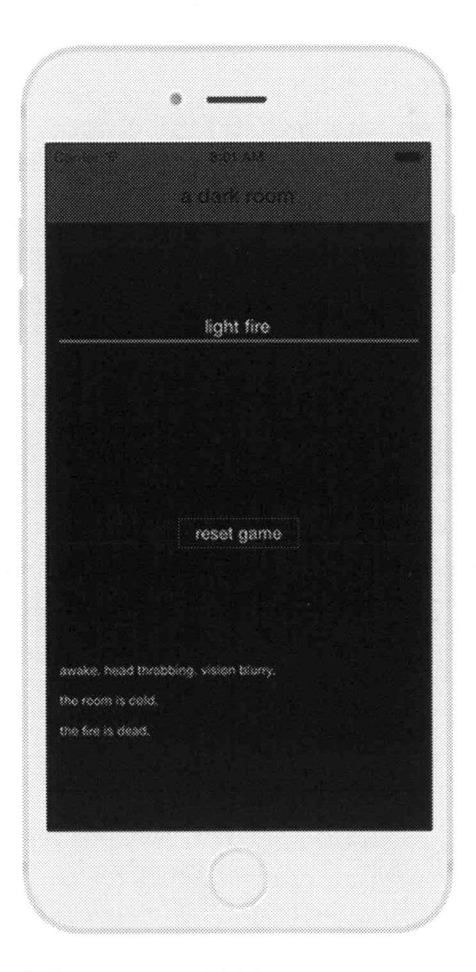

Figure 7-1. *A Dark Room app, 2016*

- *UI/User Interface*: This is exactly what it sounds like—the buttons, screens, and interface that the users use when navigating the app.
- *Wireframe*: These are the sketches of apps you've seen made on graph paper, with pencil and pen and markers.
- *Mockups*: These are the photographs of apps made in Adobe Photoshop or Illustrator before they are programmed; they simply represent what the app is supposed to look like when it is completed.
- *Prototype*: This is a test version of the app, where some screens are linked together and you can tap through the app, but it is not actually made out of real code. It's the difference between walking down the street with a cardboard cutout of a car and making engine noises, and really driving a car. Prototypes are used to communicate the idea of an app quickly and to test user workflow.

Now you'll know what these words mean when we use them, or when you see them on the Internet. You are officially a beginner UX designer.

Wait, What Is a UX Designer?

A user experience designer designs things so that they are pleasant, excellent, simple, and understandable. A UX designer finds out where you expect the next button to be, and they place it there before you've even used the app. User experience designers have a romantic responsibility; they weave the experience for the users and fulfill their expectations (or fail to) before they even touch an app.

A UX designer is not a graphic designer, as a graphic designer makes your buttons catch the eye and picks colors and pixels. A lot of software designers play the role of both.

Now that you know what user experience design really means, let's go over what it looks like in mobile design. We will go over how the average user behaves, so you keep in mind who you are designing for while you are designing. When you look at your app's menu, it will be laden with all sorts of details you notice and lovingly made—the bevels, the button size, the way the transition interacts with elements of both screens, and the meticulous care put into the app.

When the users look at the app, they see a bunch of buttons. More precisely, they'll see only the button they need to get what they want done. Your users' focus will be on the world around them.

Figure 7-2 shows what you see versus what your user sees.

Figure 7-2. *What you see versus what your user sees*

They are looking at it conceptually, taking subconscious cues from the design to discern what they need to get done. Users use apps in brief moments of time: They're sitting in a waiting room and have 10 minutes to get something done; they're at a stoplight quickly reading a text before the light turns green; they're waiting for their friend to pick them up from their house. No user stops momentarily to admire the beauty of your app's design, because a moment is all they have.

Users also don't have the time to paw around, searching for what they need. The designer luxuriates in their design, but unfortunately they are the only one who does this. Every use involves just a couple taps, and then the users are done.

Because users only have a moment to use the app, nobody will stop to try and understand your app if it's not obvious. If users don't get it immediately, they will close the app, never to return. The only time users notice design is when something is going wrong.

So not only will users not focus on your design, you're designing *specifically* so that they don't. The best design is design that's unseen, design that allows users to slide effortlessly through the app as they are going about their day. Good design will never be complimented by anyone but a designer, because it will never be noticed by anyone but a designer. The brilliance of Apple's UX is that you never notice Apple's UX.

So your users have to understand an app immediately for everything to go right. Since they have to understand the app immediately, it should be full of symbols rather than words. Symbols communicate much more quickly than words. Take, for instance, the humble yield sign pictured in Figure 7-3.

Figure 7-3. *A yield sign*

Nobody has the chance to slow down and figure out what the yield sign means, so people have to understand while they're driving down the street. If a sign is poorly designed and doesn't communicate its purpose immediately, somebody may die. Luckily, you have much less than a human life at stake.

You may not even know off of the top of your head what a yield sign looks like, but you know what it means, so it's done its job. It does this using a symbol. The word "yield" is placed inside the symbol, but the symbol takes priority. Symbols communicate intent much more effectively than words in a pinch. Consider that the first languages of civilization were based on symbols.

Make it as simple as possible for the users to understand your app and make it through the interaction. In short: keep it simple, stupid. This is not to say your users are stupid, but that they allocate only a tiny fraction of their brain to your app at any one time, because they are busy doing other things as well.

Tip Keep the cognitive load low on your app, so that your users don't have to focus on your app any more than they want.

Wireframing

Wireframing is the act of putting pencil to paper and designing the actual app screens. No more pseudocode, outlines, or use cases, but hand-drawn pictures of how your app will really look.

There are some simple tips and tricks you can use while designing that can transform apps from middle-of-the-market designs to winning, intuitive experiences that are successful on the iOS store. These tips are used during the first phase of developing your app, putting pencil to paper. You have to sit down at your desk with a piece of paper and draw pictures of the app for people to see. You can't explain how the app works to developers; you can't explain how it works to people you're marketing to; you can't explain the app to a client; and you can't explain the app to anyone without your wireframes finished and ready to present.

Luckily, making wireframes is easy. It doesn't require highly technical knowledge or special skills, just diligence and a willingness to go over your drafts again and again. You'll be spending a fair amount of time bending over paper-and-pencil sketches of your app.

Why can't you jump straight to the computer for design? Beginning your app design on a computer is going to lead you to subconsciously create something that would work well on a computer, because of the interface you are working with. Wireframes that are hacked together on a computer will reflect the mouse-control metaphor that birthed them, whereas a drawn design will most closely reflect that process of manipulating things with your hands directly. When you draw with paper and pencil, you are working with an interface that is closest to the touchscreen nature of the iPhone, and your design will reflect that.

It is not entirely necessary to start with pencil, because many good apps exist for wireframing that allow you to do it on an iPad or iPhone, the native environment for an app. The advantage of these apps is that they allow you to test your app prototype directly on the device, the best place of all to test.

Remember, wireframes are not meant to demonstrate how beautiful your app will be; wireframes by skilled, experienced designers will be beautiful in their own right, but your wireframes will and should look like a sketch. They will be messy, drawn over, redone, and redone again.

Here are some great tools for wireframing apps easily and quickly:

- *iOS Sketch Paper*: This is free sketch paper for designing iOS apps, so you can pencil in your designs on a to-scale paper iOS device before beginning to test or build them. This will save you the time of using a ruler to make boxes the same size as your iPhone. This paper is also made specifically to go with this book `https://gumroad.com/l/RgXqW`.

- *Prototyping on Paper*: Once you've drawn all your screens, you can use Prototyping on Paper to build a full, working prototype of your app on your device out of those sketches. If you're going to put pencil to paper, this is the tool to use to put it all together.
- *App Cooker*: This is a full prototyping iPad app that helps you design apps as an indie designer or a team. In other words, an app to make an app! In addition to prototyping, it has a pricing tool, mock iTunes store, icon creator, interactive wireframing tools, and a million other things. It even has a companion app, called App Taster, that you can use to test the prototypes built on App Cooker. App Taster is free, so you can share these prototypes with anyone you'd like.
- *UIStensils*: This is a metal stencil you can buy for iPhone and iPad (and other devices) so you can put pencil to paper for your designs and have them look great before you even hit the computer. Similar to Sketch Paper, it'll save you the time of using a ruler to measure out boxes for your sketches. However, they are not to scale.

Once you've drawn your designs, build a prototype. Use one of the services mentioned here to build this prototype and share it with five people. Don't bother sharing it with more than five people, but test each major change with five different people. The feedback from these people on this initial prototype is crucial, and you cannot do without it. However, chances are that your prototype testers are not user experience designers, and so do not have technical feedback to give you.

Therefore, you need to "translate" their feedback to get to the root of the problem. Here are some possible meanings of common feedback:

- The app is tough to use: That means the design is poor. Screens don't flow right or where the user's hand goes is not where the buttons are. The fix is to adjust button location to places that people's fingers naturally go.
- Reporting a feature as missing that isn't missing: You need to make this feature more prominent by making the menu items clear. If that isn't possible without compromising your design, you might need to cut the feature.
- Testers keep accidentally doing something undesirable: One button is at a location the user habitually hits (because of other apps), but the button does not do what the user thinks it should do. Ask the users what they meant to do instead.

Considering Price

One of the things you need to consider while designing your app is the pricing model. Hopefully you've already made a decision about how you're going to make money from your app, so you can incorporate it gracefully into your design. The following sections discuss how each pricing model affects the design of your app.

In-App Purchases

In-app purchase interfaces are designed by you, not automatically built into the app. That means it's totally up to you to make them work well, and the interface of the in-app purchase needs to be designed with as much care as any part of your app. They should be as natural as any other part of the app. The users should not feel as if making an in-app purchase is an awkward and painful experience. You don't want it to be difficult for people to give you their money, after all. Insert the menu in a location where the user would need the extended features you're offering.

Advertisements

Advertisements need to be integrated into the app just as gracefully or you run the admittedly high risk of alienating users due to ugly ads. Advertisements can be full screen pop-up ads or bars that run along the top or bottom of the app screen. Make this choice depending on what the user is doing. If your app is a game or immersive experience, a pop-up that appears after sections of the content is over is better than a banner ad that visually interrupts their experience.

Make sure that the pop-up ad is timed after the immersive experience is over, as studies show that video game aggression is really due to frustration, not violence. Make sure your app doesn't frustrate users, and don't interrupt serious or consequential interactions in your app.

Conversely, a utility app might be better off using a permanent banner on the screen for ads, as these apps don't provide an immersive experience. The user is only popping in and out of the app for a moment, and there's no time for a banner ad to annoy a users who are already in a rush. Opt for the permanent banner, which will always be seen no matter how little time users spend on the app.

Flat Fee

An app that charges a flat fee needs no additional interface, since it's handled entirely by Apple. Apps with a flat pricing model need no additional design consideration.

Fragmentation In Design

Apple devices now come in quite the selection of different screen sizes. This is known as UI fragmentation, meaning that the Apple UI has been fragmented into different sizes. The wireframing and design principles we just went over apply to every Apple mobile device of every size. However, different screen sizes each present their own unique challenges.

The screen layout across all of the iPhones must be identical, as well as with the iPad layouts. So you need to consider the challenges of every device you're designing for, even though you're only designing for one screen layout. Check out the *Human Interface Guidelines* for more information.

Full-Size iPad

Way long ago, in the yesteryear of 2010 when the iPad was just about to come out, many a joke was made about how it was a giant iPhone and wouldn't do well. Why would we need a giant iPhone? We already had the regular size iPhone, and that was good enough for us.

Similarly, a lot of people thought designing for the iPad would be like designing for a giant iPhone. Sure, the screen is bigger, but the principles are the same—put the header here, make sure it has the right buttons in the right spots, and you're good to go.

Designing for the iPad is entirely different than designing for the iPhone. The design principles are different, the layout is different—everything is different. The iPad is a lengthy experience, a surface to bend to your will, displaying and creating content at the tap of a finger. An iPhone is a brief foray into your content from which you quickly emerge, while an iPad is an extended visit into your favorite books, movies, and other content.

What this means is that when designing for the iPad, you need to keep in mind that users are going to stay inside your content and continually interact with it. People don't just pull out an iPad for five minutes to check something like they do with an iPhone. Design for a deep, interactive experience consuming content with the iPad.

Even the two different orientations of an iPad, landscape and portrait, are markedly different from each other. To see this, all you need to do is pull out an iPad and look at the Mail app. In landscape, the side column's view of emails is stuck there, and it isn't an option to make it go away. In portrait, the mail navigator is a toolbar, only to be seen when you don't want to navigate emails using the included arrows. The visual focus remains entirely on the document at hand.

Portrait mode allows you to consume large amounts of text in a streamlined way and get through work without breaking the flow. Landscape mode lets you see the hierarchy in a more technical way, allowing you to sift through data as you please. Portrait mode is a metaphor for consuming text and documents in one go, and landscape is a metaphor for consuming data on a computer's landscape screen.

These metaphors are true of all of Apple's native iPad apps. iBooks is always displayed in portrait mode, and when you display it in landscape the text is split into columns to read like a typical book, with two portrait views. Since the computer monitor is in landscape and is associated with productivity and data, many business and IT apps out there are locked to landscape mode or at least are advertised in landscape.

Content-creating apps fall in landscape mode. Pages, Keynote, or anything you'd associate with content creation is advertised in landscape mode. The features are reduced and the buttons are few, but in order to leave typing and content creation intact, the app launches in landscape mode.

Portrait mode was designed for reading documents, and landscape mode is for writing them. Content-creation apps don't have much action in portrait mode, because that's not the metaphor we're used to. Most languages are read left to right, top to bottom (there are exceptions). Tools for content creation can't be at the end of how we read the screen—this affords them less importance.

In a painting app, this could be a critical mistake. If the tools are clunky to reach for and it seems unnatural, that's a tick or three off of your design scoreboard. If it seems natural for your app to be exclusively landscape or exclusively portrait, don't be afraid to make it so.

The accelerometer is a nice feature and it seems intellectually like something you must take advantage of, but chances are if you're making a text editor, users won't miss portrait mode.

The ideal orientation for your iPad app should come to mind fairly easily, but the bet is still out about what you should do about the other orientation. Since Apple rejects poorly-designed apps, the best thing to do is leave your app without the other orientation unless you can do it well. Apple's *Human Interface Guidelines* would rather you lock your orientation for better design, and the App Store is Apple's universe. If you do decide to support both orientations, support them both well.

iPad Mini

These design metaphors do continue into the iPad Mini. Although the screen size is more portable (and everything is the proportion of an iPhone's screen size), it is evident that the metaphors continue because of the way an iPad mini's screen is laid out.

However, what is different about the iPad mini is the way it is used. People use the iPad mini with one hand, those with larger hands easily carrying it around. Using two hands, you can hold it up and type using both of your thumbs, eliminating a lot of the awkward resting of the iPad on pillows, bags, and laps.

Minis fit better inside of small bags and purses while out on the go, yet are not as restrictively small as a phone. This means you get the mobile device wherever you are—people do bring their full-size iPads places, but not everywhere, due to its size. It is reasonable to put a mini inside of a bag and bring it everywhere.

This is especially convenient for public transport, when one hand can be used holding a guardrail and the other used holding the mini up to watch a YouTube video or read an email.

The iPad is also especially convenient for maps and navigation, being bigger and more appropriate than a phone, while a full-size iPad is hilariously cumbersome.

Tip Make sure your interface is big and simple enough that iPad mini users aren't playing peck-and-hunt for buttons.

---

iPhone Plus Models

The iPhone 6+ presents a unique challenge to app design, because you can no longer navigate an app using just one hand. Some people need a sure grip on the 6+ with two hands in order to reach buttons across the screen. This makes button placement on your apps all the more significant. Important functions should be placed along the right side of the screen, because most people are right-handed and will be picking up their phone with their right hands. This means only right-side buttons are reachable with one hand.

A larger screen is not permission to add more buttons. With more screen real estate, it is tempting, but the use cases of the iPhone 6+ remain the same as all its smaller younger brothers, save for some extra movie watching. It's still an on-the-go device and should be treated as such, restricting functionality to only what users need in that moment.

The iPhone 6+ is the same aspect ratio as the iPhone 5. What this means is that all of the apps made for an iPhone 5 will work on an iPhone 6+. That said, they don't look great on an iPhone 6+. Like blowing up a small pixelated picture, iPhone 4 and 5 apps on the iPhone 6+ usually look poor and old. Make sure your graphics are updated for the iPhone 6+.

iPhone Smaller Models

Just as with the iPhone 6+, you should update your graphics for the iPhone 6 as well. The screens get blown up on an iPhone 6, although the change isn't as glaring (and unattractive) due to the smaller screen size. The iPhone 6 should also get its own set of graphics.

The iPhone 6 is slightly larger than the iPhone 5, so only those with small hands have a tough time reaching all of the way across the screen. This means that, like with smaller devices, everything needs to be easily readable on the smaller screen. It may be more like the iPhone 6+ or more like the iPhone 5, depending on the size of your users' hands.

iPhone 5/5S

Designing for the iPhone 5 is essentially designing for the iPhone 4, but with some extra screen real estate on the top/bottom. The fact that the width is the same makes it easy to transition iPhone 4 designs into iPhone 5 designs, as you usually only need only extra whitespace. Leave whitespace or room for scrolling content intentionally so the design is backward-compatible.

The major difference with the iPhone 5 as compared to prior versions is that, when viewing an app with the keyboard, the area without the keyboard is larger than the area with the keyboard. This is a huge advantage in text-based apps like note-taking ones. Since the iPhone aspect ratio is preserved for later devices, this advantages persists all of the way through the iPhone 6+.

iPhone 4 and Previous

Designing for the older iPhone sizes can be quite difficult, because of how small their screens are. We're getting more and more used to *phablets*, so we forget the principles of designing for such a small screen. It all boils down to one simple rule; don't add more than you need.

Since screen real estate is so precious on smaller, older devices, you need to make sure that that tiny screen can deliver to the users everything they need to know from that screen and move on. This means eliminating the detritus of things that users already knows about, such as extra text and complicated visual elements.

In the next chapter, we'll pull these concepts together by taking a look at an app called Scribe.

CHAPTER 8

Case Study: Scribe

Scribe is an app that solved one very small but very serious problem for some people—copying and pasting between their Macs and iOS devices.

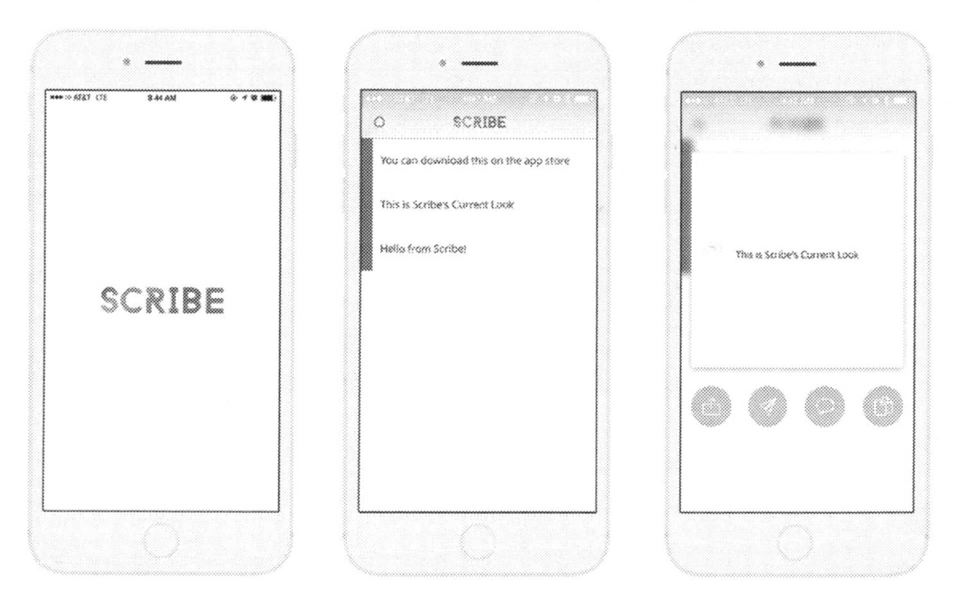

Figure 8-1. *Scribe app 2016*

M. Holstein, *iPhone App Design for Entrepreneurs*,
https://doi.org/10.1007/978-1-4842-4285-8_8

This was more of a design challenge than it seems. The team at Hipolabs, the firm that created Scribe, had to find a way to get data from the iPhone onto the Scribe app easily (and get around all of Apple's design constraints) and then get that data to the Mac, and then to make it easy for the user to copy the data wherever they want on their computers.

Hipolabs decided to go with a keyboard shortcut on the Mac, which is not as convenient as you'd think; with so many keyboard shortcuts available, finding one that is both convenient and unused was quite a challenge for Hipolabs.

What made this fantastic design possible at Hipolabs was having continuous access to their designer. "One of the partners at Hipo is our UI designer," says Serkan Terek, Strategy Lead at Hipolabs. "We looked for a long time to find the right person... he's a great guy, has a great vision and sense of minimality in his work." Hipolabs refused to compromise until they found the guy who they knew would do a good job.

Furthermore, they had their designer in the same room with them, which made a huge difference in the development process. "It definitely was not done in a single iteration," says Serkan, laughing. "Since we had the luxury to have the designer in the same room with us the whole time, we had the opportunity to play around with a lot of different approaches." You may not necessarily have the designer right in the same room with you, but constant and continuous communication is key in the design process. The time design takes can be increased dramatically if both parties are not immediately responsive whenever the other reaches out.

Don't be afraid to try a couple of wildly different approaches before settling on the one that you will use. "[Our designer] would be inspired to do something, do a draft and show it to us, where we all gave our input. It went through quite a few changes to get to the simple-looking Scribe we have now," says Serkan on the iteration process at Hipolabs. Sometimes (most of the time) you need to try several different designs before settling on the one that is right for you.

Scribe went through a major design switch before settling on the current iteration. "After spending some time with the old version, we realized how off-key it was from what we envisioned Scribe to be and switched the color scheme completely into the version we use now."

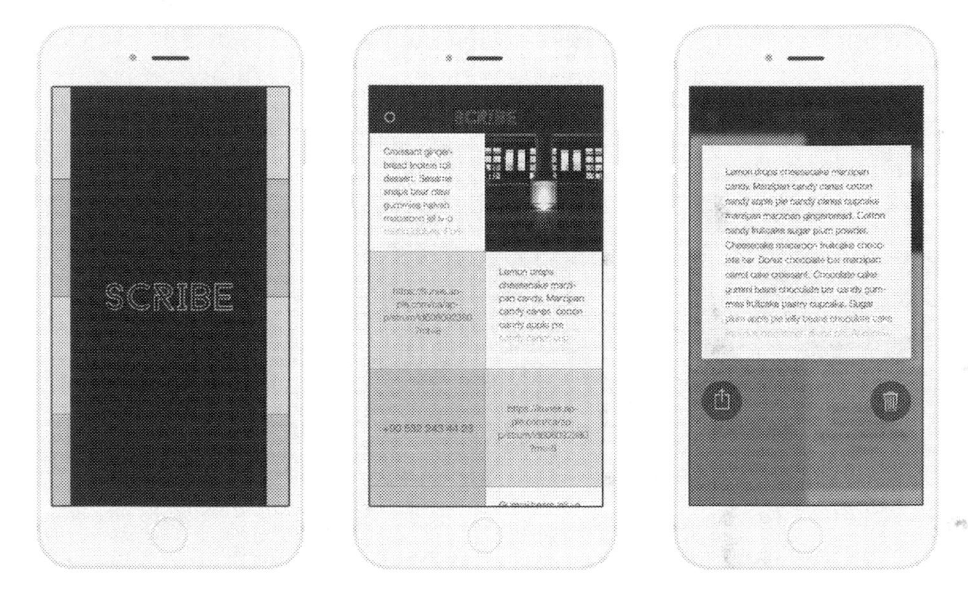

Figure 8-2. *An old version of Scribe HipoLabs experimented with*

However, even with all of this fantastic design, getting noticed wasn't a piece of cake for Scribe. To market Scribe before release, Hipolabs did research on outlets which wrote things similar to Scribe, and then sent a personalized email to each news outlet they found. "Personalized emails, I believe, were key to reaching out to the people that mattered," says Serkan. What mattered for them was not so much the amount of emails sent out, but the quality of those emails. "Your total reach out and your language makes all the difference," according to Hipolabs.

What made the difference in Scribe getting noticed was an interview in *The Verge,* after a lot of time spent pushing press and marketing. "On the day of the launch, we knew we were going to be reviewed by a couple of small sites after reaching out to tens of the biggest ones. We were kinda disappointed to be honest," says Serkan of Hipolabs. "Then, about 45 minutes before Scribe went into the App Store, we got an email back from *The Verge,* asking us whether it was okay if they wrote something about Scribe. It was rewarding, since we spent most of our time reaching out to their authors," Serkan says. Hipolabs invested their time into reaching out to big media news outlets, and it paid off. You can never get that larger coverage if you don't try, and you never know when your big break will come.

Marketing at Hipolabs did involve some dollars invested. "The few places we paid money were the high-end personal tech blogs of known authors," says Serkan "We got featured by a couple of them with paid advertising, but those were worth it, since they have one of the most devout follower bases." Reaching out to places with active readership can be better than one of large readership, because the active reader will share content and continue discussing it, making each impression that much more worth it.

The next chapter, which is the last chapter in this section, discusses how to give your app its own personality.

CHAPTER 9

Design Personality

Graphic design is the part of app design everyone is familiar with—the area in which the pixels, colors, animation, and visuals of the app appear. This is where your wireframes become mockups. Before you start on your graphic design, you need to have in mind what you want your graphic design to be like.

To figure out what you want out of the graphic design, answer the following questions:

- What feeling do you want to give your target market? (Do you want them to feel entertained, focused, hardworking, etc.?)
- What colors do you have in mind, if any?
- What are some other apps you want yours to look and feel like? (They don't need to do the same thing).
- What is the size of your app's personality? Do you want it to be part of the users' normal workflow, or for it to be big, distinctive, and entertaining? An app with a small personality would be blue and white, plain, with Helvetica Neue font. A big personality could use Comic Sans, torn-paper-effects, and pastel bright colors.

M. Holstein, *iPhone App Design for Entrepreneurs*,
https://doi.org/10.1007/978-1-4842-4285-8_9

- Consider what your app does when considering its personality size. Does your app require all or only some of the users' attention?

Tip Determine your apps personality and dress it accordingly.

The following sections discuss examples of apps that are designed well, both graphically and in user experience.

Drafts 4

Drafts is designed to quickly and easily capture text. This text is then saved and shared wherever you'd like it to be.

This is the perfect app to have a muted personality. In an app where the user is generating or recording content, the focus should generally be on the content, not on the app itself. Draft's clean interface and simple controls allow its users to focus entirely on the content that they're generating. It's the perfect app for a simple interface. See Figure 9-1.

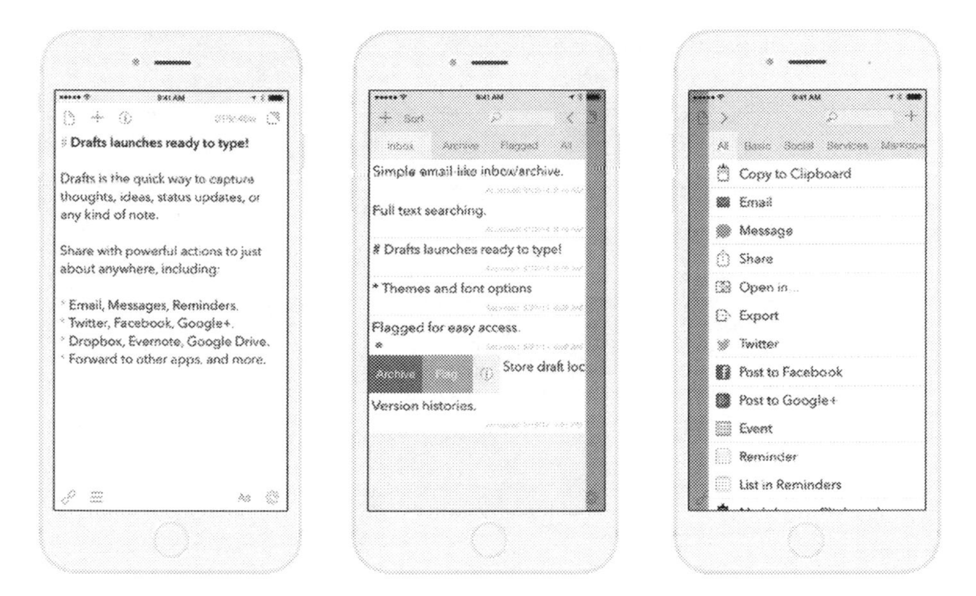

Figure 9-1. *Drafts app 2016*

Pocketbooth

One app with a big personality is Pocketbooth, a cute app that is a photo booth in your pocket. It does nothing but take photos and share them with people. A simple app, but it does something entertaining with personality that a lot of people consider hipster, which is how their primary user base identifies.

You can see that they definitely considered this when creating the graphic design for the app. They knew their primary user base was young people with big personalities (and egos), so they made their app bright, bold, and exciting. See Figure 9-2.

Figure 9-2. *Pocketbooth App 2016*

The type of app determines the strength of its personality, but every app has one. Even Apple's standard mail app has a personality of business and productivity, even if it is a personality of standing entirely out of your way. Your app's personality will be a big part of your success, and you should choose it early and remember to make it as obvious as a neon sign in your design (but not as gaudy as one).

Your app's personality should mirror what the users perceive their own personality to be—or more accurately, what they *want* it to be. If your users want to be seen as edgy tech gurus, a flat design will work best for them. If your users want to be seen as cute or fun, bright pastels will work the best. If your app is a visual content-sharing platform, neutral colors like cream or white let the content do the work for you.

Seek Critique and Feedback

There are communities out there dedicated to critiquing and developing the visual appearance of an app, starting with the first draft of wireframes and working all the way to the final mockups. Their input can be invaluable when assessing your design or a contractor's design for an app. They can be found in the same corners of the Internet you've been scouring, with the addition of Behance.

Be Creative

Note While being taught to drive, my father always said that the lines on the road were really only suggestions. If I was stuck in traffic or couldn't proceed without going across a line I wasn't supposed to, I could just go over that line. Because, really, it wasn't as if the paint minded.

Consider the app design rules we've discussed here as lines on a road. If you need to do something different or better and you know you're making the right decision, drive over the lines. If you're ever feeling nervous about driving over the lines, remember this: Good or even great user experience is achieved by following the rules and making it as predictable as possible for your users; excellent, famous user experience is achieved by breaking the rules and remaking them from your ideas.

Design is art, and art is a free and creative expression of yourself. Everything you design is a piece of you, as art is to the artist. Don't apologize for it, don't make excuses for yourself, and believe in it as much as you believe in yourself. The App Store doesn't want to crush you, and neither does Apple.

All the design rules in this book aren't meant to crush your creative spirit. They're meant to keep you from making beginner mistakes due to thoughtlessness and lack of knowledge. Now that you know about these guidelines, if you go against them, it will be because you thought about it and have a better solution in mind.

Consciously going against the rules is much different than slack design due to laziness. You shouldn't ignore an instinct to go against the rules. If you need any more convincing, pull out your phone and look at the top 10 paid apps—they don't look anything like iOS's default UI. Has a top app ever looked like Apple's default UI or design guidelines?

That's the end of the design section. Next up is getting your app made.

PART II

Development

"Begin with the end in mind."

—Stephen Covey

CHAPTER 10

Funding Sources

So you've got a plan for an app, and you've determined a way to bring that app into reality. Now it's time to tally up the costs. Let's go over possible costs briefly so you can get an idea of what you're up against.

When we account for only the mandatory costs, we come up with the following:

- $100 for the iOS developer program
- $0 - $2000 for the cost of development itself
- $500 - $1500 for contracting
- $200/yr for the app builder
- $0 for a do-it-yourself approach

These are the lowest possible estimates, and you're still looking at a total in the thousands of dollars. You need to find this money to make your app. Where can you get this investment money?

If you or a close friend/family member have the money to front and are willing to do so, that is awesome. Paying out of your own pocket is almost always the best way to go. However, not all of us have the money to make our own apps. Every method of funding (outside of self-financing) has an obligation attached to it. It is best to go through ways to finance your app development, and assess the pros and cons of each, and their corresponding obligation.

M. Holstein, *iPhone App Design for Entrepreneurs*,
https://doi.org/10.1007/978-1-4842-4285-8_10

Kickstarter

One of the most obvious ways to raise money, a Kickstarter campaign is a very popular way to test a concept and generate an email list before your app's release.

Pros

- The obligation in this method is to deliver your app to your backers. That great, because that's what you're going to do anyway! No extra hoops to jump through.
- The very act of financing your app this way gets you more users for your app. Kill two birds with one fundraising campaign.
- The ability to say that you successfully Kickstarted an app holds a lot of sway. The phrase "successfully Kickstarted" has its own weight and serves as its own accolade.

Cons

- It's not guaranteed success. You may go through all the effort to run a Kickstarter and still fail. Even if you get 95% of the way there, if you're not at 100%, you fail.
- They're hard. Not everyone attempts a Kickstarter because they're very difficult to pull off. People aren't just waiting in the wings to give you money; you have to hit the streets and sell, sell, sell.

- There's a floor and a ceiling to fundraising; amounts under $3,000 and over $1,000,000 aren't always well suited to their platform.

How to Get Started

Kickstarters seem easy in concept; create a video and an article, list some rewards, and kickstart it. But if that's all you do, your Kickstarter campaign will not be successful. For every potato salad63 campaign, there are hundreds of serious campaigns that died in obscurity. Kickstarter campaigns are hard, just like an election campaign, fundraising campaign, or any other type of campaign. They require months of work and do not come together by accident.

With a Kickstarter campaign, you're trying to convince people to give you money. There are three types of people who support Kickstarter campaigns, known as the three Fs—Family, Friends, and Fools.

Family and Friends

- Family and friends, contrary to popular opinion, are not just willing to throw money at loved ones when they simply say "I am building an app!" They want proof that this is something you're actually doing, not just something you're thinking about doing.
- Send them an email explaining why you're doing this. Focus on your personal motivations, dreams, or goals. This demonstrates that you've thought this desire through completely.

Fools

- People on Kickstarter want personal benefit for participating in your Kickstarter. Unlike with friends and family, the warm feeling of supporting someone isn't enough.
- Make the rewards for contribution compelling, and make sure one of the rewards is an actual copy of the app.

Kickstarter relies heavily on people sharing campaigns through their social networks. This means you need to leverage your entire network, and essentially spam people with your Kickstarter. Knowing how to use social media and having a plan setup is crucial to your Kickstarter campaign.

There is a right way to set up and execute this plan, and it alone could fill a whole book. If you're interested in doing a Kickstarter, do some lengthy additional research on what separates successful from unsuccessful Kickstarters, so that yours isn't a bust.

There are other websites that follow the Kickstarter model, including but not limited to:

- IndieGoGo
- Fundable

Applits

Applits is a unique service that allows you to develop an app without having to do the bulk of the app development work. You submit your idea to a month-long competition and compete with other ideas to be that month's winner. The winner gets their app developed for them, from the graphical design to the programming to listing it on the App Store.

Applits provides all the funding for development, which is why this is a funding source. You win, and they build it all for you without charging you anything. You get 15% of royalties on top of that, so it's an excellent low-risk, low-energy play. Chapter 11 includes a full case study of Applits.

Pros

- It removes a lot of the pain of making an app, because you're not managing or dealing with the actual production of your app; it is being made for you.
- It's great for people who have no dreams of managing or developing apps full time, and would be happy with a little effortless money on the side.
- Even those who don't win get a wonderful chance to get feedback and exposure to new ideas for their own app, as people comment on and improve ideas through the voting process. This is excellent idea validation.

Cons

- The downside is that you make far less money on each sale. Winners get 15% of the revenue each month. 15% of $1,000 is only $150, so if you have an app that makes $1,000 a month, which is well above average, you're still left with little more than pocket change.

How to Get Started

1. Head over to Applits and sign in.
2. Submit your idea.

That's it! You're entered into the Applits competition.

Investors (Venture Capitalists)

A more traditional way people finance a startup or small business (when it's in tech or high-growth industries) is by seeking funding from a venture capital firm or business angel.

A venture capital firm is a group of people who invest in startups with a mutual fund, and a business angel is a person who does so out of their personal account. They perform the same financial function, although the culture around venture capitalists versus angels differs greatly.

You set up a meeting with the investors and pitch your app idea or business to the group. These pitches include conventional metrics such as including sample financial earnings, assessing the market, and explaining how much money you're asking for as an investment.

Traditionally these groups do not invest less than $20,000 in one venture due to the decreased return on investment, but if you have an idea that legitimately requires more capital, you should consider this investment type.

The money is actually a trade for company equity, which means if your app is acquired or becomes huge and your company is sold, they get a portion of the proceeds.

Pros

- Excellent for people who need large amounts of investment capital, $10,000 and up.
- You don't just get access to a big sum of money; you get access to the personal expertise of the fund managers, and you get access to the expertise of other companies in the fund's portfolio.
- It is quite an accomplishment to be running a venture-backed startup. It's the "big leagues" of tech startups.

Cons

- It's a huge commitment. With this sort of investment, you're expected to drop everything else in your career and make this startup your full-time job.
- You need an app that has a real potential to make back a large sum of money. This means your financial projections can't be based on your hopes and dreams, but need to be based on hard market data and any pre-launch excitement you've generated.

Not only is this not the route for every app developer, it's not the route for most app developers. If you don't entertain dreams of being the next Mark Zuckerberg, this is not the financing route for you. If you do entertain dreams of being the next Zuckerberg, soldier on.

How to Get Started

There are plenty of websites where you can network online with venture capitalists and angel investors. Here are some:

- *Fundable*: Fundable has an interesting model. You can solicit investment money from venture capitalists and angels by creating a campaign on their website, similar to a Kickstarter campaign model. However, instead of soliciting contributions in small amounts from lots of people, you're soliciting investment agreements in large amounts from a small number of people.
- *AngelList*: AngelList is a website that allows you to list your startup on their website and indicate that you are looking for investment funding. Listing your startup on their platform and soliciting investment is free.

Additionally, you can also post jobs for your startup on AngelList and build your team through their website.

- *Google Local Firms*: Search Google for venture capitalist firms that are located in the nearest metropolitan area and you can pitch your idea directly.

Find a Business Partner

An additional partner who is a source of money may not bring as much time or skill to the company, but will provide the bankrolling necessary to earn their equity. There are many resources on the Internet that connect people with ideas to people who want to invest in an idea by joining a company.

Pros

- You get the benefit of a new partner's experience, which can be substantial.
- They may bring a fair sum of money to the table, so you don't have to involve any financial firms for funding.

Cons

- This person might be a complete stranger. A small business is a very difficult and personal experience, and going through something so life-changing with a complete stranger might be more difficult in the co-pilot seat.

- Traditionally, the more money someone brings to a venture, the less they are expected to work. If you don't bring any money, you'll be responsible for a sizable portion of making the venture work.
- They may leave the venture early or flake out.

Make sure you get quality recommendations to verify that they have good character and are trustworthy people to make an app with.

How to Get Started

There are several social networks that exist expressly to help would-be founders find other people to cofound a company with.

- FounderDating
- Founder2be
- CoFoundersLab

Check out one of these websites if you're interested in finding a business partner.

The next chapter takes a more detailed look at Applits, the company you learned about in this chapter that helps people get their app ideas off the ground.

CHAPTER 11

Case Study: Applits

Applits has kindly taken the time to write a case study for the readers of this book.

Applits is a company that relies on crowd-sourcing to bring its app ideas to market. The Applits platform takes a unique approach to the industry, running as a monthly competition where site members submit their app ideas and vote on their favorites, and the most popular ones are developed by Applits. Submitters of the winning app idea receive 15% of the profits the app makes once on the market.

Additionally, Applits members are able to win smaller amounts of revenue by participating in decision-making tasks such as naming apps, adding features suggestions, and doing competition research.

The founders of Applits—Josh Tucker and Keith Shields—envisioned Applits as a way for everyone with great app ideas to have a shot at getting those app ideas developed. Josh and Keith noticed that barriers to entry for bringing an app to market included not having enough time, the high price for app development, and lack of coding knowledge. These barriers are things that most people with busy lives are not able to overcome. After all, work, husbands and wives, and kids tend to take precedence over an idea on the back of a napkin.

M. Holstein, *iPhone App Design for Entrepreneurs*,
https://doi.org/10.1007/978-1-4842-4285-8_11

Within two years, Applits evolved multiple times. Site features have been added, analytic tools developed, and anti-cheating countermeasures introduced. All of these features are aimed at doing two things; first, they are used to keep the competition fair. Second, and more importantly, they are used to generate the most accurate market research possible for app ideas in the contest.

With hundreds of app ideas submitted every month, and thousands of votes, Applits has a great deal of market research accomplished before an app is ever put into development. Since the competition is a popularity contest, the best ideas naturally get the most amount of votes, and therefore are the ideas that are developed. By choosing which apps to develop based on the desire of the crowd, Applits has a unique ability to gauge what apps smartphone end users want.

What is so clever about the Applits platform is that they already know the apps they are developing will sell well, because the ideas won their competition. They've integrated idea validation into their platform.

After site members choose an idea to develop, they choose the name, slogan, and features list for the idea. Once that's taken care of, Applits works closely with its community to perfect design and development of the app. When Applits introduces a new design or development update, the Applits community is there to offer feedback in the way of suggestions, praise, or criticism. Applits then takes this feedback and implements it into the app. Once implemented, the updated app is once again shared with the community, and the cycle continues.

Figure 11-1 shows one of the Applits apps, called Umbrella.

Figure 11-1. *The Umbrella app, 2016*

Applits has built an entire platform idea validation during the app development process. However, the good news is that you don't need to run a company like Applits to get great feedback and market validation on your app idea. Here are some quick ways to get feedback on your app idea:

- *Attend an event like Startup Weekend:* Events like Startup Weekend and CodeDay71 are outstanding ways to get feedback on your idea. Not only do you get the opportunity to meet lots of people who can help get your app idea to market, but you also get to do an early test to see if people are interested in your idea. If people are interested in your idea, it gets built. Applits did this to validate their company idea, and users do it to validate their app ideas.

- *Start a survey:* In your survey, make sure to ask questions about demographics to gauge who will use your app, how much they would be willing to pay for it, and what features they would like to see in the app. The important thing is to make sure you're not only asking your mom and friends for feedback, since they will be too nice. Go out in the real world and ask people to fill out your survey whom you may not know in everyday life. Part of the competition process at Applits involves community members answering questions like this during the competition.

- *Try to sell your product before it's built:* This might sound stupid. You don't even have a product yet! If you have a good idea, go try to sell the concept to people. If they want to buy it and ask for the link, you know you should probably go make that product a reality.

Unless you're talking about some top-notch intellectual property that's truly unique to your product (like engineering a new type of wind turbine), sharing your idea to get feedback on it is only going to amount to great feedback. Most entrepreneurs aren't out there to steal other's ideas because they are either too busy working on their own company, or they have other ideas in their head that they'd rather be working on in the unlikely event they ever have any free time.

Figure 11-2 shows another Applits app, called Draw my Life.

Figure 11-2. *The Draw My Life app, 2016*

The Importance of Constant Feedback

During design, make sure to reach out to as many people as you can to see what they think. Attending events at your local tech accelerator is a great way to find people who can offer feedback. People are often super helpful and eager to see you succeed. Make sure they can effortlessly understand what each of the buttons on your app does and how to navigate it, without having to do much hard thinking.

Finally, you have to worry about development. There are plenty of things to keep on eye on here, and feedback—once again—is crucial. A product that Applits relies on heavily is TestFlight. TestFlight allows developers to easily share pre-launch app builds to beta-testers. Oh, and it's free!

Applits finds beta-testers in the group of people who voted on their app, in their community. What Applits does hinges on having an active and engaged community. You can use a community in the same way.

Applits thinks that giving the end users a say in what apps are made is great, and has the added benefit of collecting some great market research before the product goes live. You can't always blindly listen to feedback.

The next chapter discusses the different ways you can get your app made: do-it-yourself, app builders, and hiring someone else to do it for you.

CHAPTER 12

Development Methods

You have your idea all fleshed out, and you're ready to get your app made. However, there are a couple of options available as to how you can get your app made. Each has its pros and cons, and each costs varying amounts of money.

The three basic options are:

- Program the app yourself.
- Use an app-building or app-making program.
- Find someone who knows how to program.

Lets take a deeper look at all of these options, to see which is the best for you.

Do It Yourself

Building the app yourself is exactly what it sounds like—learning how to do programming, graphic design, or both, and building your app all by yourself. This is the clichéd image of working nights and weekends to build something on your computer in the dark and secret, until you unveil it to the world.

M. Holstein, *iPhone App Design for Entrepreneurs*,
https://doi.org/10.1007/978-1-4842-4285-8_12

Pros

- This method can cost as little as $100, as your only mandatory cost will be a developer's license.
- You have complete control over your app and graphic design. Your exact vision is built, nothing more or less. You can make whatever you can code and design.
- Programming is an in-demand skill. You can take your programming skill and the app you've just made and use that as resume fodder for a job or a move into a new career field. Having a published app on your portfolio looks really good to potential employers.
- The ability to design user interfaces is also an in-demand skill, since every piece of software needs someone to do the graphic design. UX is a lucrative and unsaturated career field.

Cons

- It's more financially efficient to hire people to do it for you (unless you're already a programmer) because you can make more money doing stuff you already know how to do.
- You shouldn't teach yourself to program if you don't like it. Programming is a very cool and respected skill in today's workforce, but you'll never find success forcing yourself to do something you don't like.
- It takes anywhere from three months to a year to make your app if you're teaching yourself, because each new feature presents an entire new set of things to learn. It

is a little like trying to build a house from the ground up when you know nothing about construction.

- If your app becomes successful, you'll have to hire additional programmers, because you'll need to focus on business/managerial things.
- If you're interested in programming and want to sharpen your skills with an iOS app, build it yourself.

So how do you decide? If you want to make an app and see programming as a frustrating barrier, you're better off choosing another method. If you want to code the app yourself because you want to be cool, just know that people are always impressed with those that have made their own apps—no matter how it got done.

How to Get Started (Programming)

Here are some resources for those who would like to teach themselves how to program.

For beginners:

- Code Academy
- Your First iOS App

For those with some programming experience:

- Treehouse
- Pluralsight

There also exist many classes or boot camps for programming, which teach you programming in 8-10 weeks. These can be excellent places to learn, as they keep you on track and you don't get burnt out or lose focus as it goes on, due to the external structure of classes.

The big downside is that they require a full-time commitment, and if you have to support yourself or someone else, this isn't an immediately feasible route.

How to Get Started (with Graphic Design)

You can also learn do to the graphic design of your app yourself. All the benefits of doing it yourself apply to graphic design, as well as all of the drawbacks. So in summary, if you are genuinely interested in graphic design, learn it yourself. If you're not, it's better left to someone else.

Here are some programs that you can use to do your graphic design yourself:

- *Adobe Photoshop or Illustrator ($50/mo)*: These are popular tools to start off with, as they are powerful image editors that allow you a high degree of finesse with your design. Photoshop and Illustrator are very similar, but Illustrator has a slight edge because they do vector graphics—graphics that are never pixelated. Photoshop does this too, but only in certain situations. These are very technical tools, and you need to be willing to invest a realistic amount of time into learning these programs.

  Decided to go with Photoshop? Pick up the handy component cut and slice me, which makes creating and organizing the .png images for your app a snap.

- *Xara ($300)*: Xara is a competitor to Photoshop with cheaper prices—around $300 flat—and it includes graphic design and web design components.

  The drawing/illustrating component of app design programs does not matter as much as the layout capabilities, and Xara has these in spades. Many professional graphic designers have eschewed Adobe for Xara.

- *Gimp (Free)*: The greatest part about Gimp is that it is free. People call Gimp "the free Photoshop," and plenty of Gimp guides exist on the Internet. It is very similar to Photoshop and has all of the same capabilities.

  Gimp also has well documented help, and there is a wealth of tutorials on YouTube and the Internet in general on how to use it. The principal drawback of Gimp is that it is outdated; most newer Macs won't even run Gimp at all.

App-Building Program

App-builder programs enable you to build your own app. They don't require any programming knowledge to use, and they can be downloaded easily—and you can get started using them today. Think of PowerPoint, but for making apps.

Pros

- Ideal for someone who wants to build an app for their own personal amusement or for a casual side project, due to their easy-to-use nature and drag-and-drop interfaces.
- Doesn't require that you have any programming knowledge, or that you hire a programmer or partner with one to get your app made. This means that you can get started today.

Cons

Hand-coding an app (whether you're doing the coding or someone else is) can bring into reality whatever is in your imagination. With app-building programs, you're limited to what the program can do.

You're limited by your own personal talent in understanding the app program, and each of these services requires their own specialized knowledge. You may end up sinking as much time into understanding the program as you would teaching yourself to code, depending on the service.

If you use the default app elements that are available in these services, you risk the unique graphical impression of your app in the App Store. An impression that, unfortunately, can make or break your app. This means that you'll still need to hire a contractor for the graphic design or do it yourself.

Also, there's a technical consideration—as the platforms (iOS and Android) mature and age, your app will become obsolete and poorly coded ten times faster. This means much more upkeep over the long run, which (like a car) you don't think is a problem until it's a problem.

Your app will never be astronomically successful, because of the server/feature limitations. This isn't to say you won't generate some spending money, but app-building services don't lend themselves to record-setting successes because of their infrastructure. If your app does become a record-setting success, you'll probably have to hire people to re-code the app natively.

How to Get Started

Here is a partial list of app-building programs, so you can get started building your own app.

- *GameSalad* ($299/yr): GameSalad is a fantastic alternative for people who have an idea for an iOS game. They boast over 70 top-100 games, proving that their app-making platform really is up to the job.

Their engine supports various types of games (sidescroller, runner, etc.)—you just need to provide your own content. GameSalad can be a life-saver in the world of iOS games.

- *AppMakr*: AppMakr is a service that turns a website RSS feed, Twitter handle, YouTube channel, or website into its own app. This is an excellent solution for people who want to make apps for their blog, news website, or content-delivering website easily. It's perfect for YouTube channels or blogs that want their own app.

  There isn't a lot of customization, but you can "skin" the app by changing the colors and look of the app AppMakr produces. This is worth looking into if your app is an accessory of a content-based website, and you already have a strong brand.

  The concern with AppMakr is that it does not provide a unique look or feel for your app, so your app will be not much more than a mirror of your current website. Most UX professionals these days suggest that you should simply optimize your website for mobile instead, using a responsive design. AppMakr cannot produce the individuality in design or user experience needed to create a truly successful app.

Hiring Developers

Contracting is just hiring someone else to build part or all of the app for you. The price of this can vary, depending on the complexity of your app. This is how most companies build their apps; they hire a programmer to build the app for them.

Pros

- Hiring a contractor is liberating. You can get a simple app programmed for as little as $200, and the contractor will be doing what they love, and you will be doing what you love.
- If your app is even a fourth as good as it should be after all the planning you were put through, it will easily make back that small development cost in the App Store.
- You spend less time working on building the app yourself and more or all of your time doing some early marketing and advertising for your app, so that when your app launches, it can have a big launch day. The time investment of doing either or both of these things on your own far outweighs the time commitment of managing contractors making your app.
- All of the advantages of building it yourself (such as retaining complete control, freedom, and imagination) are still there.
- You'll never have to switch development methods later, and therefore your app's development will never get derailed because of this.

Cons

- The cost can be intimidating if you look at above-average industry estimates, or quotes from boutique agencies. The cost for an app created by a famous agency can be upwards of $1M, but any app can be made for under $5,000 with proper budgeting and management.

- You still have to put effort into the app, most obviously spending the money on the project. You also have to closely manage your programmers, because letting an error get away from you can quickly become a big problem if you're not managing it closely or correctly.
- It is risky. Just like paying someone to do work on your home, if things go wrong, the contractor can take off with your money and leave you with nothing.

This method of app development is—if it's done right—the act of starting your own very small company. If you ever dreamed of having your own company, this is the path for you.

How to Get Started

There are a couple of ways you can contract your app development, and each is suited best for people in different situations.

- Find individual contractors on the Internet to work with. Some websites you can use to do this are listed next.
- Find local individual contractors. You can do this through websites such as Craigslist. The advantage is an in-person contract and lowered legal fees, and the disadvantages are, of course, having to handle the legal and financial aspects all by yourself. If you do this, check out the legal app Shake to handle your contract. (Using Shake to manage distant contracts is not a good idea, because law enforcement gets more complicated with distance.) This is recommended if you hope your app will evolve into its own company someday.

- Find an agency, local or distant. Agencies are groups of people that can build your app for you. Designli.co is one example, and Teehan+Lax is another. They handle the legal and financial aspects, do a good job (if you find the right firm), and will handle all aspects of production for you. The single disadvantage is the higher price tag.

Here are some websites where you can hire freelancers:

- oDesk
- Elance
- Freelancer
- Craigslist
- 99Designs

Since Craigslist does not have legal management, you can circumvent the attorneys fees and paper contracts by using Shake. You can use it to build simple contractual agreements, like the one you make with your programmer.

Here are some agencies you can hire to make your app:

- designli.co. (To get $300 off your service, mention that you found them through this book.)
- Teehan+Lax

All three ways of making an app can lead to polished and fantastic apps if the app-maker is (you are) dedicated to making the best app possible.

Choose your method of app development based on how you want to spend your time making and your goals for the app, not because of outside peer pressure or a feeling it "has to be done this way".

The next chapter discusses the ins and outs of hiring your own developer. If you're planning on building it yourself or using an app builder, feel free to skip it.

CHAPTER 13

Hiring Developers

If you've decided to contract out your development work, you need to know how to avoid common mistakes and pitfalls when working with a contractor, whether that's locally or online.

Creating a Job Listing

The first step to contracting is to find a contractor, and this means posting a job listing. Posting an effective job listing (one that will bring in good candidates) requires a couple of steps.

Determining a Payment Method

First, you have to decide whether you will pay them a lump sum of money, or by the hour. Since you're working with 1099 contractors and not employees, both are completely acceptable methods of payment. They each come with their own advantages and disadvantages.

M. Holstein, *iPhone App Design for Entrepreneurs*,
https://doi.org/10.1007/978-1-4842-4285-8_13

Fixed-Price

Pros

- Maximum control of budget.
- Contractor is incentivized to work as efficiently as possible.
- Payment is exchanged only when you have a product in your hands.

Cons

- Budget and requirement changes need to be explicitly renegotiated each time.
- You need to be vigilant about paying your contractor so that they're receiving the money.

Hourly Contract

Pros

- Contract is more flexible; can easily add on extra budget and time to handle new issues.
- Contractor's stream of income is much more consistent.
- Contractor is incentivized to spend more time trying new things.

Cons

- It's easy for a contractor to pull one over on you by wasting time coding things they don't need to code.
- The contractor is not incentivized to be as efficient as possible.
- You pay the contractor regularly, even if they don't deliver a product.

There are many risks with hourly contracts, and they're directly linked to how little experience you have with the contracting world. Hourly contracts are more flexible, but also make it easier for the contractor to hustle you.

Hourly contract theft looks like this: They'll reel you in during the interview process, giving you a fantastically low budget and wonderful value. You, as the client, are excited about the deal you got. Then, they take as long as possible to deliver the project to you. They'll code by hand things you could have found in code libraries and include poorly implemented features you didn't ask for, and then they'll charge you to remove them as well. There will be too much code that took too long to make, because they were trying to maximize their billable hours. It's a quick way to lose all of your money if you're not diligent and reviewing their work an hour or two every day. The next thing you know, you have a shoddy product and went over your budget by as much as 100%.

On a fixed price contract, contractors are not going to exhibit behavior like this. They're intent on getting the contract finished as quickly as possible, with as little work as possible. So instead of people wasting your time and coding extra features, they're going to use pre-packaged libraries of code if you provide them and are going to write your code as efficiently as possible. They won't code any features, and they will code exactly what you asked and no more. This means algorithms that aren't unnecessarily large and code that runs quickly and does only what you need it to do.

If you're thin on cash, you could also make your payment deferred. For instance, you could pay 30% of the contract three months after the contract is over. This represents a chance for you to pay the developer back with some of the app's earnings, instead of pulling cash away from your own savings. The risk here is that even if your app doesn't make much money in the first three months, you still need to come up with payment. If you want to do this, make a note of it in the job listing.

Determining Where to List Your Job

There are many websites where you can post your job listing. Some provide full features for you to manage your contract, and some are mere job-listing websites. Consider these websites:

- *oDesk*: oDesk is a global website dedicated to facilitating the client-contractor relationship. They have a system that helps you find a candidate, hire them, work with them successfully, manage their payments and taxes, and then close out the contract. It's the one website you'll need in the contracting process.

  One of its biggest benefits is that the legal aspects (non-delivery of code, contract breach, etc.) are all handled within the website so that if the worst happens, you get quick resolution.

- *Elance*: Elance is the same as oDesk in that they are a one-stop shop for your client-contractor experience. Elance and oDesk are owned by the same company, having went through a merger recently. The difference between them and oDesk is mostly aesthetic, with vastly different interfaces having the same basic functions.

  Elance includes legal management as well, so that you don't have to go outside Elance when something goes sideways.

- *Craigslist*: Craigslist can help you quickly find applicants in your local area. It's one of the first places people think to go when looking for work, and so it's rich with people of all different skill levels.

With Craigslist, you'll need to manage the legal aspect separately, but it can be done for free through an app called Shake. This app allows you to avoid attorneys and pre-made contracts online and all the fees that come with them.

- *99Designs*: 99Designs is a website aimed at graphic design contracts only. It has a unique model; you post your job listing (called a contest) on their website, and graphic designers compete by submitting entries. This means that you will get dozens of different entries to your contest. You pick the winner, and the payment is transferred to them. They also support one-on-one projects, if you've found a designer you love and want to continue working with.

Writing Your Listing

In order to attract quality candidates, your job listing needs to say more than "Looking for iOS app programmer, email resume to companyname@gmail.com." Providing more information to potential candidates will end in a better fit and will save you a lot of interviewing time.

Things You Should Provide

Your organizational materials. This includes your wireframe, prototypes, app outlines, and descriptions of your app. Provide everything the candidate would need to know about how your app works. This is not anything new you need to create; you have created these materials over the course of developing your app idea. A prototype would be excellent.

A proposed schedule. This schedule explains the milestone intervals (such as every two weeks, every four weeks, and so on), how much should be done at each milestone, and when you want the app to be completed.

Make sure that you explicitly state that all paid milestones must be delivered in either .psd files or Xcode files (whichever is relevant), and make sure they understand that personally before offering them a job.

This prevents contractors from receiving payment, and then holding your code or graphics hostage, forcing you to pay them the full sum of money (maybe more), even when the product they developed was sub-par or nonfunctional.

Refuse to work with anyone who doesn't agree to this in text before the start of the contract, so that if the worst happens, you have evidence that they agreed to that as part of their contract.

State that you're looking for a full-time commitment out of people who apply. A full-time commitment for three weeks is quicker than a part-time commitment for eight weeks or an as-needed commitment for six months, and all are going to cost the same amount. Additionally, if they state they'll give it a full-time commitment, it often works out to a part-time commitment; part-time will work out to as-needed work; and as-needed commitments work out to a dead project.

Additionally, a full-time commitment generally means higher-quality work, since you've literally bought the contractor's full attention. That is why corporations buy cubicles, huge buildings, and office equipment for every one of their employees, when in a digital age half of their work could be done online. Corporations are trying to get their employees' attention.

State that development agencies need not apply. Some will apply anyway, but by-and-large you should not accept their offers. Agencies have several layers of communication between you and the programmer, which results in the vision for your app becoming diluted. What ends up being coded may not be what you intended.

They are also far more likely to hold your code hostage in exchange for the full sum of money, even if they did not build what you hired them to build. This is going to cost you money and time, both in dealing with the agency and in hiring someone else to clean up their mess.

Ask Questions

You should ask questions in your job listing. When they respond to your listing, they need to answer these questions in their cover letter. This ensures that they actually read and paid attention to your job listing, instead of just blasting their resume to any email they could find.

Let applicants know that getting the job done ahead of time will result in a bonus, and/or getting it done behind schedule will result in their pay being docked (if that is within your contract rules; consult the website terms and conditions to make sure).

These financial incentives can make the app development time three months when it may have taken six. Also, explicitly ask the applicants to overestimate potential schedules instead of underestimating them, so no applicant is shining you on with a ridiculously quick schedule.

Ask the contractors to explain their vision to you, if they have one. This is a way to ensure that they can adequately communicate in English, which is important or you'll lose money because of communication barriers. This also shows whether the candidate is excited about your app idea.

It also helps you ascertain whether the person is a good fit for this job; someone who has not captured the vision you have for the app will not do a good job, no matter how skilled they are or how much money they are paid. This is very subjective, so trust your gut.

Vetting Applicants

After your job posting has been online for a day or two, you'll likely have enough applications to begin vetting them. First, you need to make sure you have the right number of applicants.

What you should have after a day or two is 20-40 applications; fewer than that means that your budget is too low, and many more than that means it's too high. If you received too many or too few candidates, take your job listing down, repost it, and try again.

You need to whittle your 20-40 candidates down to just a few people. Luckily, this process goes fairly quickly.

- The first thing you can do is reject all applicants who applied to your job but didn't answer any of the questions you asked in your job listing. If those people didn't communicate with you over the first stage of the interview, they are certainly not going to communicate with you during the rest of the contract. Asking to discuss the question further in a message or chat is not a red flag, because it means they're replying to you and communicating with you, but outright ignoring the questions is a red flag.
- Reject any applicants whose estimate is significantly over- or under-budget. Over-budget, for obvious reasons, and under-budget, because estimates that far under-budget come from inexperienced or unconfident contractors.
- Reject everyone who writes very poorly in English. If their English is so poor that their formal, edited cover letter is hard to understand, their conversational English is likely to be unintelligible. This isn't a matter of racism. As much as you might like to, you can't have a productive work experience with someone whom you can't understand. (Of course, if you speak their native language, go ahead and ignore this.)

Note Don't preclude people just because their profile is new and they don't have many logged hours or they lack professional experience. People who are just breaking into working on oDesk understand that they're newer and some of them will work harder and for less money in order to quickly bolster their profiles. Some of these candidates have a lot of work experience, just none on your particular website.

If you give them more money than they asked for and promise to give them glowing feedback (if they deserve it), you will have a wonderful and inexpensive experience. This builds a good relationship, and you can easily go back to this person for app updates and expansions.

Whoever is left has made it to the next round of the vetting process. You are going to put your applicants through a couple more paces by asking them more personal questions about their work histories. These questions must be saved for your personal contact with them.

Candidates are being judged on their answers, but they are also being judged on the quality of their informal English. Not everyone writes their cover letter themselves, and it's important that you're able to communicate with someone properly in informal English.

If they don't respond promptly to these interview questions, it's likely that they won't work quickly and efficiently on your contract either, and it's likely they won't communicate with you as much as necessary. They need to be quick to respond in order to keep the momentum going.

This should narrow your pool down to two or three people who accepted all the terms of your job application and responded quickly and articulately to your follow-up interview questions. At this point, you can conduct Skype interviews and make your decision.

Listen to what your intuition says about candidates; if one candidate seems too good to be true, even if they passed all the preliminary tests, they probably are. If you think someone is the right fit for your project, listen to your intuition. If someone doesn't feel like a good fit, don't hire him. If nobody feels like a good fit, don't hire anyone.

Getting Started

So you've picked your candidate and are ready to get started. There are several things you need to iron out before you begin.

Define Succes

You should determine beforehand what "success" means for your app project. That seems a little silly at first, because it seems obvious what success is—having the app working well and on the App Store, of course. Unfortunately, a project can be a failure for a great number of reasons, even if it is on the App Store:

- It did not work as intended
- It never made back its investment and turned a profit
- It cost significantly more than planned
- It took significantly more time than planned

These are all important because you have to make plans based on these requirements. If you fail on time or budget, you may have to sacrifice on marketing or other efforts, even though you have a functional app.

The problem is that a lot of contractors don't see these as criteria for success because they are not criteria they have to consider. The contractors are not privy to your budgeting decisions, profit margins, or plans you made based on time estimates. This can range from innocent lack of knowledge to downright lack of concern (since they only care about how much money they can make off you). You can make this something they are concerned about by threatening bad reviews for going over time or budget and providing financial incentives for being on time.

Make sure, before work begins, to define what success means for this project. Let the contractor know your goals and find out what your contractor's goals are for this project. They are not without goals, and you want to do everything you can to make sure their goals are achieved.

You can attach bonuses and rewards to fulfilling all of these goals. If they fulfill all the goals with no problem, they can get more money.

You don't have to pay your contractors for one hundred percent of the work, if they do not do one hundred percent of the work. You define one hundred percent of work at the beginning of the contract, by defining what success means for your contract.

Lastly, it helps your contractors do better work. They can only do their job if they know fully what their job is; you need to take the time to explain to them exactly what it is you want them to do. This can help them achieve what they're being paid to do.

Determine Milestones

Milestones are an important part of a relationship with a contractor. A milestone is any exchange of product built by the contractor in exchange for payment in a fixed-price contract. In an hourly contract, it is any point that the contractor delivers the product.

It was mentioned briefly earlier, but you want to determine all milestones and agree on them before you begin. You want to do this so that everything about the contract is spelled out beforehand, and you aren't stuck in an argument in the middle of a contract.

There is a spectrum of how many milestones you can have, ranging from one (you give them the money and they give you the code at the end of the project) to very many. Always take more milestones over fewer when you can, for a couple of reasons.

- The more milestones there are, the more you and your programmer/graphic designer are communicating. This means that when a miscommunication happens, you catch the miscommunication—and the resulting error—before the contractor has done too much work. This saves you money and it saves both of you wasted time.

- You can give smaller chunks of money away for each milestone. This evens out your financial burden and the contractor's cash flow and keeps the amount of money proportionate to how much work the programmer has already done.

Determining when your first milestone will be due is easy. Simply ask your programmer "What can you get done in two weeks?" and then reply to whatever they answer with "All right, I'll be expecting that them." This way you know you're not rushing anyone or working too slowly, because they themselves set their work amount. It keeps everyone accountable.

When receiving a milestone, don't pay for anything but an actual .psd or code. There are versions of these files (.pngs, .jpegs for images, .ipas for apps) that can display what was created but in a non-editable way. If you were hiring someone to build you a house, this would be like paying for a photograph of the house. It's evidence of the work done, but not ownership of the work itself.

If the worst happens and the contract ends badly, you're out of luck if you didn't require the actual .psd files or Xcode project. You'll need to file a claim with the website, go to court to force the contractor to fork over your code, or live without it and start over. Don't put yourself in that position.

Keep in Contact

It is so easy for contractors and clients to lose contact and focus on the project. When you lose focus, your contractor loses focus, and then the project dies in the water because everybody got focused on other things.

You and your contractor should settle on regular meeting times (e.g., every Friday), so that you don't lose track of what's happening. You can keep in contact through:

- Status emails
- Skype call/video chats
- Phone calls

Every time you meet up, you should be looking at the app and discussing it. Deliverables are so easy to produce for mobile apps, that you can be discussing the real project and the real code at every status meeting—further keeping everyone moving forward, on track, and accountable.

Test the functionality that the contractor said will be working at the status meeting, so that you can double-check whether or not something is working. Contractors might say something is working, when it is only barely so or it is not done well and needs checking over. This often isn't done maliciously, but out of forgetfulness.

Don't be afraid to contact them. If you have a question about something or are wondering how they're doing, don't feel like you have to leave them alone or that you're bothering them. Just refrain from sending them an email more than every second or third day, which might come off as spam.

If they do fall out of contact, don't immediately assume it's because they've dropped your contract and don't care anymore. Things happen; people have family emergencies; they get suddenly busy at their full-time job—you just don't know. Just make sure to let them know that you'd like a heads-up when they will be unavailable.

Note It is a red flag if you ask them to tell you why they were suddenly gone, and they don't give you a reason. People who are invested in contracts may not always get a chance to let you know beforehand, but they'll always let you know afterward as to what held them up.

Another danger is with someone who is continually late, every time. Disasters can come up now and again, but disasters do not come up with every milestone. If your contractor is behaving this way with you, it may be time to renegotiate or end the contract.

If the relationship is not working out, don't be afraid to let people go in the middle of a project. In the long run, it is much easier to end a bad relationship halfway through and finish a project with a new, better working relationship than it is to drag the bad one out all the way until the end. The quality of the working relationship will definitely affect the outcome of your product, so don't let yourself stay with a bad contractor because it seems easier. In can be difficult to let someone go, but you'll feel better afterward.

Remember to keep your freelancer included in the conversation, as opposed to just giving them materials and telling them to "do it". They presumably have experience making apps and so may have some valuable insight to offer you. Make it clear in the beginning that you welcome contributions and feedback, because they will probably assume that you don't. Don't close off a resource that may end up being more helpful than you anticipated. (Besides, if your app becomes a huge success, you might want to keep working with the same person and build this wonderful app together. Don't damage that relationship before it has a chance to flourish.)

The next chapter discusses essential code add-ons you'll want to include in your app.

CHAPTER 14

Code Add-Ons

There are sets of code you can get for free and add to your app directly, which can save your developer time and you money when it comes to adding key features to your app. You want to know about these before you begin the development process, so that you can consciously include them in your app.

App Templates

One of the tools available to an app developer, whether they're doing it themselves or working with someone else, is an app template. An app template is a skeleton framework of code for a certain type of app—they are like stencils for apps. Templates come in all shapes and sizes, such as "sidescroller game" or "to-do app". All you have to do is make new graphics and apply them to the template, perhaps do a little code editing, and the app is made.

People use app templates because they can save a developer an extraordinary amount of time. All that time you're spending figuring out how to put together the code or algorithms for a certain function, when you can just pay a little money and use a premade version. You can also buy templates where the graphics are already done for you, and it is the codebase that needs to be written instead.

Is using an app template right for you? Let's look at the pros and cons.

M. Holstein, *iPhone App Design for Entrepreneurs*,
https://doi.org/10.1007/978-1-4842-4285-8_14

Pros

- They're designed to save time on the programmer/designer's part when making an app, much like a stencil would save an artist time when reproducing a drawing.

Cons

- What you need to be wary of if you want to create a winning app is services that simply supply the template code and let you change the graphics a little, and then resell. Remember the ideas and the features that make your app idea are different, and you put your hard work into making those features stand out. You don't want to create just another app using a template-to-app service.
- Using templates has a legal risk. They will usually license code use instead of selling full rights, and if you ever want to sell the apps or incorporate them into a business model—basically, if you ever want to experience huge success —you won't own the code that made you a success, meaning you'd have to code the app from scratch all over again when the time comes. Sometimes it's more important to get to market and you can go back and fix it if you're a success, but be aware of what you are doing.

To demonstrate why these services are not the best route for an innovative, successful app, just think: Have any of the best apps you've ever used been so standard as to fit inside a template? You can easily make another to-do app, another iPhone scroller game, another note-taking app, or typical apps with these templates pretty easily.

However, because of their nature as reproducible template services, no new innovative or disruptive app is going to come from these templates. They may start from one or borrow the code from one, but they are going to be modified to suit the original innovative idea. If you want your app to shine as much as possible in the marketplace and to be different and better than the millions of competing apps, shy away from template-based services and their reproducible results.

If you do find an appropriate template for your app, by all means use it. You can then add all the custom work necessary to make it unique, and then release it to the App Store. However, this means you'll have to find an app template for which you will have full rights, and not just licensing rights.

How To Get Started

Here are a couple of websites where you can purchase app templates:

- *AppTopia*: Their templates are already fully built and just require different graphics. They are already optimized with advertisements, so you can get your app on the App Store and making money immediately.
- *MyAppTemplates*: They templates give you the opposite of what AppTopia does. They have fully built graphics and a skeleton UI, but allow the programmers to build the app's insides however they prefer.
- *Premium App Templates*: They provide the same assets as MyAppTemplates, with the .psd files and icons ready to go for you.

App Store Analytics

There is a lot of important data that the App Store won't collect automatically for you. Apple will tell you how many people downloaded your app and how much money you made, but that's about it. Luckily, the free market has created a lot of solutions in the form of web apps that allow you to track everything you need about your App Store business to make good decisions.

Some services sift through all the data Apple provides you about the App Store so that you can make sense of it. They expand on the iTunes connect interface so that you have tons more useful information.

For example, AppAnnie Analytics uses just the information Apple provides and organizes it all so you can view it by downloads and revenue, across all different sorts of criteria like currency, country, and time. You can analyze your competition, your app rankings, and sales projections through their portals.

They provide a daily/weekly email digest (your choice) about your app performance on the App Store. This way, you can keep up with your app's performance, even when you're too busy to take the time to check on it every day. They'll also let you know when your app has been featured or is ranking well on the App Store through these daily digests.

Internal Analytics

There are packages of code that your developer can embed inside the app to give you usage information from inside the app. They are premade and easy to install. These include critical metrics like:

- How long people use the app
- Which screens they're on most
- Which buttons they press the most

Aside from being really fun to look at, this information can be used to identify:

- Slow or confusing screens in the app (when they remain on one screen for too long)
- Parts of the app that users don't want (screens and buttons with no interaction)
- Parts of the app that users love (when they go back and forth between related screens quickly)

If people aren't using a certain part of your app at all, you must either figure out what's wrong and fix it or remove it entirely. If people use a certain part of your app a lot, you know to go ahead and make it even more awesome for your users.

You want these analytics packages installed as soon as possible, even before you begin your beta-testing, so that you can take advantage of the bug-tracking information and analytics immediately.

Some services that provide these app analytics include:

- *Google Mobile Analytics*: If you like Google's Website Analytics, this is the perfect pair. It's accessed through the same interface as their web analytics, and has well documented installation. Like Google Analytics, it is also free.
- *Flurry Analytics*: Flurry's base plan is free. Flurry is a popular analytics option for mobile developers. They have demographics estimations based on real, hard data, as well as providing industry benchmarks for your app and comparing your app to them, to see how you're doing against the competition. They provide a more expansive solution than Google's, with greater support.

- *Localytics*: Localytics is free up to 10,000 active monthly users, which is a lot harder to achieve than it seems. This is a good analytics solution for an app that's not free. Localytics is a great product because they include marketing features with their analytics, such as A/B testing, push notifications into your app, and in-app messaging triggered by events. For instance, if your user takes too long on the "edit" screen, you could prompt them with some more information on how to use it, thus providing the information they need, when they need it.

The next chapter, which is the last one in the development section, discusses how to get your new app on people's devices to test it.

CHAPTER 15

Beta-Testing

Beta-testing is a critical stage of app development that a lot of indie developers don't spend enough time on. Many indies build their app and they test it within their personal network and perhaps get their friends to test the app, but don't test the app with target users.

Mostly, this error leads to some usability challenges for the user. Function x and y are not easy to use, which is the difference between a four- or five-star review. That's not good, but it's not crippling either.

However, if you don't test thoroughly, a hidden bug can ruin your reviews. For example, consider a bug on a settings screen that happens only under certain circumstances but causes the app to delete all of the user's data from the session. These are the kinds of bugs that lead people to angrily leave one-star reviews on your iTunes page, and those are crippling.

Reviews don't go away once you update your app; reviews continue to affect your app reviews' average for the rest of its lifetime. And like your grades in college, if you get all Cs your first semester, you're going to spend the next three years trying to drag that grade point average up. It's even worse with apps—with apps, there's no graduation finish line.

M. Holstein, *iPhone App Design for Entrepreneurs*,
https://doi.org/10.1007/978-1-4842-4285-8_15

How to Beta-Test Right

In order to avoid this poor outcome, make sure the people beta-testing the app are real potential users. Better yet, beta-test the app in real user circumstances.

Is it an app to help Minecraft gamers have information when they need it? Have them play Minecraft and use the app when they need information. Is it an app to help people develop characters while writing a fiction novel? Have them use the app to develop their characters while actually writing their novel.

This may all seem obvious, but it is meant to illustrate that what you should not do is just hand the app to testers and say "use it," which is what people end up doing. You may know you should really beta-test with actual users, but fail to do so because you just want to get through the due diligence and release your app, which is what you're really excited about.

There is a problem with the deliberate testing style of just saying "use it," which is that beta-testers are aware they're testing. People treat an app differently if they know they're testing it, and they treat it differently if they know you're watching. People need to test the app the way they'd use it; in the spare two seconds before they can turn left on the road, sitting on the subway from work to home, or lounging on the couch watching TV.

You could just give them the app to use this way, but you can't watch your users use the app. Then, you can only receive commentary, and unless your beta team is a team of designers, chances are slim you'll receive great design and use feedback. The feedback will be all bug-oriented. Make sure to watch some people use the app in real-time, to draw conclusions by yourself. Compromise; spend some time watching others use your app and give the app to others to test on their own.

When letting people beta-test your app, do not help them if they don't understand something. If they can't understand a gesture in the app, only help them out when otherwise they'll have given up. This tells you where the pain point is for your app, and you can fix these by providing new controls and gestures or repairing bugs.

A fine thing to look at is what gestures they try to use, regardless of whether they actually exist yet. The user isn't wrong when they use an unsupported gesture; you are wrong (as a designer) not to include it, unless you consciously objected.

This is also good for seeing those little issues that you're missing, not just for identifying a bug. You need to see people interacting with your app in the world, and since you usually can't go creeping around convenience stores and watching people, this will have to do.

It's really difficult to get your app on many potential users' devices, because you have to find these users and then put the beta app on their device. There exist free and paid services to get both of these jobs done for you.

Finding Beta-Testers

Finding beta-testers presents yet another opportunity for you to increase the audience before the app is released. If you find people who are interested in your app and get to beta-test for free, then they will download your app when it is released.

Additionally, people who beta-test like doing so because they get to be part of something from the ground floor up and participate in the community. These sorts of active and engaged users are a wonderful asset to your growing app company, and they can turn into raving fans.

Here are some places you can look for those fans-to-be:

- *Erlibird*: Erlibird is a website where people with beta software list their software to look for testers. It's free to list your beta on their platform, although they do offer sponsored spots on their website.
- *Museum of Modern Betas*: MoMB is a blog that posts about betas. They're very informal, just short listings, and you can drop an email to them to have your beta considered for posting.

- The Beta Family: The Beta Family is a website where you list your app beta on their platform, and similar to Erlibird, users find your listing. They have a premium option where you get access to quality testers (ones who will actually use the app a lot and write long reviews), but their base service is free to use.

  The Beta Family also has a built-in system for delivering the app beta to your beta-testers, saving you the additional work of managing a separate system. However, this means you can only serve beta-testers you find through The Beta Family.

- *StartupLi.st*: StartupLi.st is a blog-style website similar to MoMB that lists products currently in beta. Make an account with Twitter and post a short listing about your app startup to reach people interested in beta-testing your app.

Keep in mind these websites tend to have a limited audience of high-tech power-users, and if your target market isn't young and technical people, you may have difficulty finding your audience this way.

There are also other places on the Internet where you can find people interested in beta-testing software. You can post on relevant forums or subreddits asking people if they're interested in beta-testing software or find other forums with a relevant userbase.

You can also reach out to people who signed up for your website's waitlist/newsletter or followed your app page on social media to solicit beta-testers. You've been building a waitlist for some time now, and it should have a healthy amount of entries.

Incentivize people to beta-test your app. Always incentivize people; don't just expect them to do something for nothing. Easy incentives include giving away your app for free to people who beta-test it or providing them some sort of free benefit. You can also give them a free download of content that they would enjoy.

Getting Your App on People's Devices

Once you've found users who are willing to beta-test your app, you need to get your app on their device. You could invite them all over to your place and install the app using an iPhone cable, but if you've collected many beta-testers in your market like you have, that will be unfeasible.

Tons of services exist to help you remotely install your app on beta-testers' devices. Remote app installation like this is great because your app will be on the tester's device whenever they want to use it, and they'll experience it like they would any other app. You can get the app on users devices, even if they're across the globe, and collect feedback about the app through built-in analytics (mentioned earlier).

Make it easy on these beta-testers; don't require that they give any active feedback. Instead, learn your way around the in-app analytics and depend on them. However, remember to follow up with your testers and ask for optional feedback.

Here are some services to get your beta on people's devices over the internet:

- *HockeyKit*: HockeyKit is a free option that is available on GitHub. This can be a great option if you'd like to modify their code and get it to do what you want, and they have the best free bundled services of any of these options. The drawback is that you need to know how to use the GitHub code yourself, or your programmer has to install it for you. If you really like HockeyKit, they have a premium package at HockeyApp.

- *Applause*: Applause is a high-end premium testing service. They cover a range of testing options, testing every functionality in the app and collecting all kinds of data. It's extremely easy to get started, and they have fantastic service in case you have any problems.

They are even available for every type of mobile app platform. That said, this premium service comes with a premium price tag. To get a pricing estimate, head over to their website.

- *DeployGate*: DeployGate positions itself as the simplest to use beta-testing service. You upload your app and share a URL with your beta-testers, and they download from that URL. They also make it incredibly easy to install their SDK. They have a free plan, but it does not include in-app statistics. Additionally, you can only test up to four apps with 20 people on the free plan, but that should be enough for you.
- *TestFlight*: TestFlight is the big kid on the block. They are recommended by Apple and are free at any usage level. The interface can be tricky to learn, but a free SDK and free beta-testing (with as many apps and users as you want) is a great deal.

That's the end of the development section. The next section, and the last, discusses getting your app on the App Store and into the hands of buyers.

PART III

Deployment

"Action is the foundational key to all success."

—Pablo Picasso

CHAPTER 16

Pre-Launch Marketing

Before you get started marketing your app, you have to do to first things first. This means making a plan and getting all your ducks in a row before launching your marketing plan.

The first part of this is setting an official launch date. This gives you a date to plan all of your marketing around. Make this date roughly one and a half months after your app itself is complete; this gives you time to deal with any problems that crop up, and gives you time to make a marketing plan.

After you've set a release date, submit your app to the App Store for review as soon as possible. This gives you as much time as possible to deal with any problems Apple may have with your app.

When you submit an app to the App Store, you're asked when you'd like to release it. Make sure to select "release when I approve," as opposed to "automatically release," and choose the date you set as your release date. You should do this for a couple of reasons:

- This means you can go in and change the Availability and Release dates to be the same date, which is the date you picked for release. This makes your app show up as a "New App" in the App Store listings, putting it right on the front page of your category. If you don't do this, it will be much farther back in the listings.

M. Holstein, *iPhone App Design for Entrepreneurs*,
https://doi.org/10.1007/978-1-4842-4285-8_16

- You need time to research, develop, and execute a marketing plan that will get you downloads on the day of release. This includes all of the subsequent chapters in this book.
- This gives you time to get your App Store listing right, so that when your app is released, nothing is left on the table or needs corrected.
- You've got time to verify that every inch of your website and app listing is crisp and clean, and that it looks exactly the way you want it to. If it automatically releases and there's an error that you didn't catch, you can't go back in time and fix it.

Essentially, you want to give yourself time to prepare for this. You only release your app for the first time once.

Your app will not market itself. You don't put your app on the App Store, have it accepted, and then watch the downloads roll in. If you sit back and wait for the downloads, the downloads will trickle in like a shallow creek in the middle of summer, and you'll wonder where you went wrong.

One of the biggest things that gets in the way of an app selling is silly errors like broken links, misspelled words, and out-of-date websites. When you control your release date, you have time to make sure everything is complete when the day comes for your app to come out, and then you can enjoy your release day knowing everything was taken care of.

The next chapter discusses how to set up your App Store listing so that the most people possible download your app.

CHAPTER 17

App Store Listing

The most important marketing for any app is listing the app in the App Store. Every listing is formatted by Apple automatically, so the only thing that differentiates your app is the content you put on it. This is the thing everyone in the world sees before downloading your app.

Here are some guidelines for getting the customer to that screen, and making every moment count when they're there.

Keywords

Keywords are the most misunderstood part of the Apple iTunes listing. Apple's search engine doesn't search the description, only the title and the keywords, so keywords are coveted space in the iTunes listing. People devote hours upon hours trying to demystifying how Apple searches keywords, and to this day Apple will share with no one how their algorithm works.

The only thing that Apple has shared is that in-app purchase names are treated just like app names, so we know in-app purchase names are indexed. This leaves it up to us to figure out how keywords work on the App Store, and the following sections describe what the community has settled on.

M. Holstein, *iPhone App Design for Entrepreneurs*,
https://doi.org/10.1007/978-1-4842-4285-8_17

Updates

Apple doesn't seem to care when you updated your app because they know that as long as it's compatible, users definitely won't care when an app is updated. This means older apps are just as likely to hit the top of the search as brand-new apps. This merely increases the pressure to choose keywords better.

Singular or Plural Words

You only have enough space for one more word, and you don't know if you should give it the singular or the plural form. Go with the plural form, as Apple will bring up both singular and plural results for the search of a plural. Using singular will give you a slight disadvantage because of this recent adjustment on Apple's part.

Articles

Words such as "a," "the," and "and" are ignored in Apple's search engine, and so you should leave them out of your keywords as well. It may look grammatically ridiculous, but keywords are only read by computers, not people. Leaving articles out decreases the number of characters you used, leaving more characters available for other keywords.

Publisher Name

Apple doesn't search for publisher names, and so if you have a strong brand, add your iTunes publisher name to your keywords. However, for the indie developer it is better to leave it out, as very few people search for apps by publisher.

Keyword Specificity

Everybody wants to use single, broad words in their keywords to catch more people, but since everyone has this idea, it is a bad idea. If one of your keywords is "love" or "kids," chances are that thousands of other people are using that keyword, and you're competing with all of them for a spot in the search results. However, if someone searches for "kids math" and you have that specific phrase in your keywords, you'll show up a lot higher in the results. It's better to be higher in the results for fewer searches, than in the middle of 50-100 apps for many searches, where nobody scrolls down to look.

You can use Google's Keyword Planner to decide on search phrases, because search patterns on Google and on the App Store mirror each other enough for the Keyword Planner tool to be useful.

Copyright Infringement

Hopefully this doesn't need to be said, but do not use any copyrighted words you do not own the rights to in the keyword field. This is strictly illegal, and it could get your app removed from the App Store and banned forever, making that app's code, name, and so on, forever useless.

Itunes Listing

Icon

This is the only visual provided about your app, so above all it needs to look very clean and professional, befitting of your app's category. The symbol also needs to be descriptive of what the app is or does, but not so descriptive it loses its beauty. Design the kind of icon you would be attracted to.

Name

To take advantage of naming, some people craft long names for their app in the App Store. This makes the app name have far more keywords.

The more keywords in your app name the better, but make sure your actual app name is easy to pronounce and can go viral—all the things a name should be.

Screenshot

The most prominent element of your app's listing on iTunes, the one most looked at by users, is the screenshot section. The most important element of these screenshots is the very first one, which shows up automatically when viewing an App Store listing on any device.

Since screenshots are so important, it's imperative that they're clean looking. Some guidelines for clean screenshots are:

- Leave the promotional bubbles and discount pictures out. Not only is this against the iTunes guidelines, it's just bad looking.
- No stickers. If you do add some text, make sure it's nice looking, not gaudy like a grocery store discount.
- Use every available screenshot area. Even if your app has less than five main screens, you can do settings screens, loading screens, or change options for each screenshot. You could frame your app in all devices, or list extra features. Utilize every screenshot slot you have.

There are two ways to use your screenshot section, as displayed in Figure 17-1.

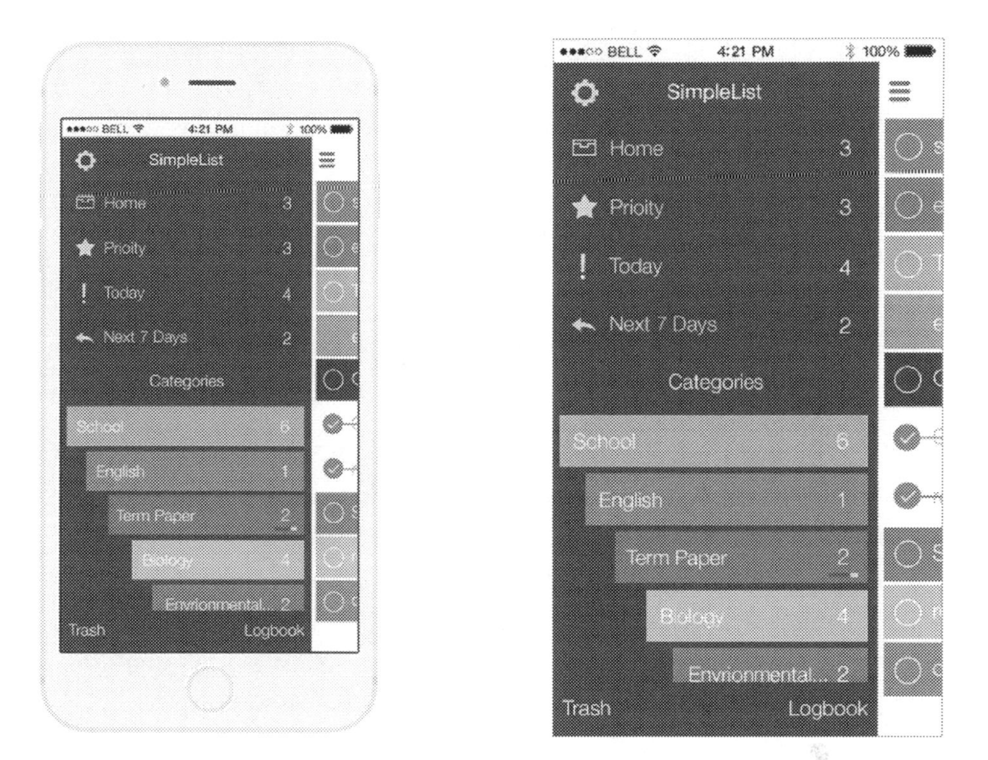

***Figure 17-1.** A framed versus unframed screenshot of the app SimpleList, 2016*

The same screenshot, framed versus unframed.

Frame Your Screenshots

Place your screenshots in a frame of the most recent iOS device. This has been more of a recent favorite and can really class up the way your app looks. This isn't recommended if your app is graphically busy; it works better with apps with a flat design.

- This can be done using Photoshop, which can give it an altered look (although still crisp and clean).
- It can be done in real life, by taking a photo of someone holding an iPhone. If the photo is taken well, this can look really good, but it's hard to catch a good photo.

Leave Them Unframed

Leave them as they are. This way, people can open them up to full screen on their devices and see what the app would really be like. You also know this will look nice and professional, no matter what your user is using to look at your app.

Description

Your App Store description is where you get to brag your app, and where people will read about it and learn more before making the decision to buy. Make that decision easy for them and provide them with the information they need (and no more).

This information shouldn't be just sales copy. It can be sales copy at the top, but as the description goes on, you should provide tangible information about features and specifications. This allows nitpicky users to make sure your app has what they want, and enables people to check your app's details out before they buy.

On the computer, only the top of an app description is always visible whether or not the user selects it, so this sentence must be something catchy. You only have one line to work with, which doesn't give you a lot. Some ways to use this line are as follows:

- One option is to tout positive reviews and press in the bullet points at the top of your list. This can be done with a variety of text characters (°? / ▸ / · / *). It sends the message to people that others have used your app and liked it, taking away the uncertainty of being a trailblazer.
- Another option is to put an app slogan or catchy phrase in that visible one-line space. This is something you can do even if you have no press or reviews yet.

After they click Read More, the highest thing in the description should be a two- or three- sentence description of the app, keeping things simple. People reading here still need to be sold on your app. You could use this space to answer:

- Who is this app designed for?
- What is the app's purpose?

One of the most frustrating things in the world is having to read a solid block of marketing for an app when looking for information. Because of this, paragraph format is highly discouraged, as people tend to skim over walls of text, and you don't want them skimming over your description.

List format works much better, because people like lists. When you see a list, you're much more inclined to read on. It improves the organization of information and makes it much easier to process. In short, people are lazy about things like this, and you should cater to that.

Guidelines or a small How-To should be pushed to the bottom of your app description. People won't want to read it unless they had the prior motivation, and if they have the prior motivation, they are willing to scroll to the bottom of the description to get the information they need. If they have the motivation and can't find the information, they can easily become angry users who leave bad reviews or don't buy.

Additionally, put support and contact information at the very bottom of your app description, so that if people look in the description for contact information instead of the sidebar, they can still reach you. You can also use this spacc to list any known bugs or bugs mentioned in the reviews and assure potential buyers that you are working to address the issue.

Ratings and Reviews

Ask everyone who beta-tested or reviewed your app prior to release to get on iTunes and review your app immediately when it comes out. An app in the new releases section that already has five-star reviews has a huge advantage, as most people check out reviews of an app before they purchase.

Good ratings are important because people consider them highly, and it's really easy to get a bad rating. There are a lot of people who will download a free app and then review it poorly because it isn't what they expected or needed, or because they experienced a small bug. Users have high standards for apps, even given their price.

Additionally, most people won't download a paid app unless it has a repertoire of good reviews, so having good reviews out of the gate is critical. One of the things that can affect your review quality is poor customer service. People put a lot of stock into how well a seller treats them; treat them well and they will reward you with word-of-mouth marketing and a good review (if you prompt them). Treat them badly, and they almost surely will leave a bad review and send other people away from your apps.

If you can, always reply immediately to emails and phone calls over your customer support. An immediate reply doesn't have to be anything more than "We got your email and would love to talk further about resolving this issue." People simply want to know their contact was received. Don't make your user jump through hoops to get in contact with you—be there to help them, because that's what you were doing when you decided to make an app in the first place. If users try to get in contact with you, they should succeed immediately.

As a general rule, the secret to success is not by focusing on "sell, sell, sell!", it's by helping people. Like happiness, money will follow if you don't focus on capturing it. Be sure to make your user's life better, and don't give anyone a reason to give you a bad review, and you will see your app turn a profit quickly.

The next chapter discusses how you can use email lists and newsletters to effectively sell your app.

CHAPTER 18

Mailing List

I would be remiss to leave out the importance of your email list when discussing marketing techniques. Your email list is an asset we're going to be using repeatedly throughout the rest of the marketing section, so you need to make a good one now.

Email lists not meant to be used to tell people to download. People opt in to your newsletter because they believe you'll email them things they want, not a stream of ads.

We're going to go over two types of mailing lists you are going to use in your marketing: your opt-in mailing list and your press mailing list.

Opt-In Mailing List

An opt-in mailing list, the first type, is comprised of people who indicated interest in your company and decided to sign up for your mailing list. This is the only thing these folks have in common; they may not be press, in your industry, or anything else, but they're interested in what you have to say.

An opt-in mailing list is infinitely better than lists of random people you contacted, because you know they want to hear from you. They indicated interest, and you won't be wasting your time on people who don't want to hear it, but investing it in people who want more of what you do. (Additionally, sending out "cold" email blasts—emailing a bunch of random people—is illegal.)

M. Holstein, *iPhone App Design for Entrepreneurs*,
https://doi.org/10.1007/978-1-4842-4285-8_18

To build an opt-in mailing list, follow these couple of easy steps:

- Make sure you have an opt-in field on your website's home page. You should already have one from validating your idea earlier. This can be big and obvious, or just at the top of the website, but make sure it is obvious to visitors.
- You can do this using Sumo. Sumo is something you can install easily on your website and set up all different sorts of options for email collection. This can be done with any website.
- Add an invitation to sign up for the mailing list at the bottom of all your customer support emails. That way, happy users can get more involved with a company that helped them out. This can read as "Sign up for our mailing list," "If you're happy with our service, consider joining our mailing list," or any variation you prefer.
- You can incentivize joining by offering people free content if they sign up. They sign up for your mailing list, and then their free content is available to download.
- For instance, users could receive a free five-step guide for starting a healthy diet if your app is health or exercise related.

Press Mailing List

The second type of mailing list is your press mailing list. This is comprised of news outlets, blogs, review websites, and any media outlet with a connection to your audience. You will use this as a way to easily contact media and press when marketing your app.

The larger this list is, the better, as it is the people your app company contacts when you have any news or updates worth mentioning. It's something you're going to be constantly building as you find new websites and blogs that would be interested in your app.

Most people's first instinct seems to be to find out the contact information of large websites such as TechCrunch, VentureBeat, Huffington Post and others, and then email them all about the upcoming release of their app several days before launch. This is not what you want to be doing.

First of all, you haven't given them any reason to be interested, just an annoying email from a company they've never heard of before. Give media a reason to care about your app being released. Secondly, going after a few big websites might not be your best play. Targeting many smaller, specialized websites is a much better first step for you.

A Few, Big Websites

Their audience is not as motivated to buy your app, since your app is less likely to be relevant to them. This is because their audience is not targeted.

Unless you're targeting startup valley boys, you're not likely to get sales. It's extremely difficult to get their attention, since they have such a large audience. Coverage is more about prestige and pride than actual sales.

Many Smaller Websites

On small websites, the audiences are very active and they will engage with you about your app. Their audience is very focused, so you're reaching actual potential customers.

It's easier to get the attention of a smaller website, so you are more likely to get coverage with any given email. You can build a lasting relationship with these outlets, building community and social capital.

You can still add large websites to your mailing list, but they should be far outnumbered by smaller websites. After getting covered by many smaller websites, you can use this momentum to get covered by larger websites. Large media outlets are much more compelled to do a story about your app if many other websites loved your app as well.

So, where can you find these blogs and websites to add to your press list? It's easy to list big websites, but not as easy to find smaller ones.

Don't worry about contacting these websites now; you're going to get all of their URLs and email addresses. The type of email you send them will depend on what you want to say, which will be covered in subsequent sections.

- *Start with BuzzSumo*: They index articles on the Internet by tons of different metrics, and you can find those targeted, impactful websites by searching your keywords on this website. When an article comes up, go to the website it's from and add their contact information to your mailing list.
- *Download this book's marketing directory*: It is a contact book for app blogs and media. Take the contact information from any relevant websites and add it to your personal press list.
- *Search your keywords on Facebook and click on the best-looking results*: Then, get their contact information from their external website. Some of the pages won't be connected to external websites, so take down their Facebook email address as contact information. When you email that Facebook email address, it sends a message to their Facebook page.
- *Do the same on Twitter*: If they have an external website link, get the contact information from that website.

> Unfortunately, if they have no external website, you have to send them a tweet asking how to get in contact. If they reply to you, take down their contact information.

As people review your app or promote it on their websites, you can add these websites to your press mailing list for future contact as well. This way, you are constantly growing your press mailing list.

You are not going to be able to find all small-medium websites in your niche today, or even in the next year. Over time, you'll find more and more of these websites as your app's presence and success grows, and as you find them you will add them to your press list. Your press list should be always growing.

At first, 20 or 30 websites on this list will do quite nicely. However, the more you can grow this list, the more return you'll see on various types of marketing campaigns.

Choosing an Email Service

So you've got email lists, and now you need to send them all emails. You need to pick a service to send emails, so you can send multiple copies of the same email quickly and manage all of the replies.

- *MailChimp*: MailChimp, like so many other services in this book, has a wonderful and full-featured free plan. Their service is free up to 2,000 emails, which can take quite a bit of time to build up to. Additionally, they have free image hosting and a very easy-to-use campaign builder. MailChimp also allows you to publish blog posts automatically using your RSS feed.
- *Aweber*: Aweber isn't as visually appealing as MailChimp, nor is it free, but it makes up for it in a

ridiculous amount of features. For people who want powerful features (like autoresponding and campaign tracking) and large lists, Aweber is the right solution. Aweber also allows you to push recent blog posts automatically by using your blog's RSS feed, which can save a lot of time in scheduling email blasts.

- *Constant Contact*: Constant Contact is the third in the trifecta of email delivery services. Constant Contact has a time-based free trial for you to test their services, which is free for 60 days.

Writing a Good Email

Note This information only applies to your opt-in mailing list. Your press mailing list is going to be contacted much less frequently, and only on certain marketing occasions.

You do not need to be a great writer, or even write your own content, in order to have a good mailing list. You can write blog articles and posts for your company, but you can also repost others articles (with proper citation, of course) from your industry. Anything that would be useful to your users is great for your mailing list.

These blog posts or articles can even be used to attract more users, as a newsletter with helpful information gets forwarded onto other friends and family members, and those friends and family members become newsletter subscribers and eventually, users. You want to make sure your email is as catchy and shareable as possible.

Here are some guidelines on writing a good email.

Subject Line

Subject lines perform best when they follow a couple of simple rules. Headlines should get people thinking about the subject (and opening the email for more info), not thinking about whether or not they should open the email.

- Keep it less than five syllables
- Use the words "you" and "your" in the subject line
- Make the headline a list wherever possible

Keep the Message Focused

Make sure to edit your email so that it stays on focus. You do not want any sentences that branch off-topic, or are much more wordy than they need to be. Don't make the user read more than they have to to get information out of your email.

This is why lists and listicles are so popular; list formats keep readers focused on the information therein, and nothing else. They expose the important information and strip away needless sentences. Infographics are even better for this.

Avoid Attachments

Corporate email servers are notorious for rejecting email attachments, so don't attach anything that isn't crucial. This means your email cannot have any extra attachments for design or branding that appear in every email.

Proofread

Make sure to glance over your whole email before sending it, to catch any glaring or destructive errors that are made. You don't want to have to send out a second newsletter, correcting a mistake you made in the first.

Check all links in your emails. Broken links are very frustrating, and they keep users from doing the thing that you wanted them to do—going to the website.

Make sure your email makes sense and is easy to read. You want this newsletter to hold value for its readers, not to nonsensically push its message. This means you shouldn't undervalue good design in your newsletter either, so choose something with photos and a font that is easy to read.

Tip Don't let a mistake go out in a live emails that could have been prevented.

Respond Promptly

This doesn't have to do strictly with your email, but a prompt reply to any contact you initiate will look great to the person you contacted.

More importantly, a late response to contact that you initiated will not convey good things about you. Make sure your responses are prompt.

The next chapter discusses social media marketing and how it fits into your marketing strategy.

CHAPTER 19

Social Media

You don't need this book to know you should have social media. Effectively using social media is another obvious step in the path to making your own app. The challenging part of social media is not making the accounts, but actually coming up with good content that gets you more downloads, more press, and more success.

This chapter gives you an intelligent starting point for using social media.

Good social media isn't just having social media accounts; good social media gives someone a reason to visit these pages. Namely, a helpful and useful community of people around the product.

There are two basic strategies for app developers to leverage social media, and which one you choose depends on your business model. If you chose the long tail business model, your social media strategy will revolve around your brand or developer account, as opposed to having a profile for each individual app. If you instead opted to put all your love and care into a single app, go ahead and make an individual social profile for your app.

The Flagship Strategy

If you have only one app and you're going to be pouring your attention into it, it is wiser to have one social media page per platform that is dedicated to your app (and not your company).

M. Holstein, *iPhone App Design for Entrepreneurs*,
https://doi.org/10.1007/978-1-4842-4285-8_19

"But they're called company pages. Why wouldn't I make it for my company?" People don't care what company made your app. There is no specific reason why people would care what company made your app, but there is a specific reason people would care about a company page that goes by your app name—you're building a brand around that app name.

Startups are starting to catch on to this, which is why more recent app companies incorporate with the name of their apps. Snapchat incorporated as Snapchat Inc., Facebook incorporated as Facebook Inc., Yo incorporated as Justyo.co, so on and so forth. You only begin to need corporate branding when your company begins to either get notable (as in TechCrunch front page notable) or you begin to produce multiple apps.

With the flagship strategy, for every social media post, the focus should be entirely on the your single app. The company brand doesn't matter at all, as the company brand is the app brand. If you have multiple flagship apps, apps that don't share a distinct brand, they function as multiple small companies.

For a great example of a company employing the flagship app model, look at Cultured Code. Cultured Code has been around for a couple years, and they have only ever made one piece of software, called Things. Even the Cultured Code Wikipedia page redirects to the Things Wikipedia page.

They are an excellent example of folks not caring what a company name is. Cultured Code exists to produce Things, and that is what they are hedging their bets on. All of their brand recognition lies with the Things app, and all of their hard work lies with it also.

Tip The flagship app's social media is about providing other useful information and resources to help the users solve the same problem they're trying to solve with your app.

The Long Tail Strategy

If you have a lot of apps that share a user base, having one corporate social media page for your apps is recommended over having a separate page for each app.

It's true that people don't care about the company that makes an app, but they do care about apps that can help them. If your one company has lots of apps that can help the same person, they will care what your company name is so they can find more apps like the first one.

Tip Make it obvious for these users where they can find other great apps by consolidating the company's social media accounts.

Since all your apps are similar, you're bound to be running similar promotions and update schedules. Keeping this all on one page means you only have to manage one social media campaign, instead of four or five concurrent ones. In the same vein, any giveaways or publicity you have is going to interest people who bought all of your apps, since all of your apps target the same market segment.

It's easier to produce content for a consolidated page, because instead of managing several separate accounts and their content, you only have to populate one set of accounts with content. With more apps, you're going to have an easier time coming up with content for one page.

What this means is that each post should focus on your line of apps, as opposed to an individual app itself. One post may speak about App #1 being updated, but the next will speak about press for App #3.

Your social media posts should help your users find your other apps, which will help them solve the same problem their current app does.

What to Post

So we've been over the goals for social media, depending on your business model. But what should you actually post on your social media? To answer this question, we have to start with the purpose of social media content.

It's really easy to just say "well, to sell more apps!" But, when people try to use Facebook and Twitter to directly sell apps, they are disappointed.

It takes more than 10,000, even sometimes more than 100,000, people connected to your page for a post to cause even a minor bump in sales, because of the low conversion rate of social media. If you're trying to drive sales through social media, you're going to have a tough time, because social media pages are not for driving sales.

Social media, for everyone, is designed to help you stay connected. Who does your company want to stay connected to? Your users, of course. People use social media to message their friends for gossip and to connect, not to discover products. It's a great place to create Internet relationships as a person and a company, to interact with your users in the same manner that they interact with each other.

Social media pages are for engaging fans of your app, people who have already purchased your app and liked the page. These are the people whose lives you're trying to improve, so go talk to them on social media. Listen to their problems and ask what they need help with. Engage with them about their lives.

Tip The best kinds of social media posts are ones that engage with current users.

What do these look like?

- *Questions.* Ask questions on your Facebook page, such as "What do you struggle most with in (blank)?" or "What's the biggest problem you face?" This both engages users and gets you focusing on their problem.

- *Pictures.* Pictures that evoke relevant emotions in your users will get shared, and if it has a watermark of your app or company name, that's your image being spread around the Internet. This is especially effective in areas that drive emotion, such as children's education or healthcare apps.
- *Videos.* As with pictures, videos can evoke emotion and get shared. Even sharing a video you didn't produce can get engagement up and make your posts more visible on social media. Also, on most social media, videos are programmed to be the most visible post type. If you post a video, it's more likely to be seen by someone than any other post type.
- *Feedback.* Post screenshots of the next major version or describe the features and ask your users whether they'd be interested in what you're building. The more you show, the more feedback you get.

This isn't to say you can't post about prominent press your app receives, milestones, or discounts and promotions—you can and should. It just shouldn't be 100% of your company feed, as it can accidentally end up being when you don't know what else to post.

People can fill their social media with posts about progress of the app, new releases, that sort of thing, but those are better suited to press relcases than social media. Folks aren't generally interested in hearing a live stream of your progress—they are interested in their own lives and what is relevant to *them*. So engage with them there.

People don't appreciate spam; they want to talk to real people, whether or not that person is representing a company. Balance in your posts matters— seem like a real person (or real group of people). This means being personal in your posts and willing to share emotions with your audience.

To do this, you can share relevant and helpful content, whether it is authored by you or someone else. Many companies run a separate blog and post their blog content to their social media websites. It is a great way of increasing an audience, because it gives your audience something relevant to read from your company.

Tip Perhaps you're writing an article about financial software to track your expenses. You could include your $4.99 layman's app for individual or household expense tracking, the $14.99 startup-oriented competitor, and QuickBooks for an enterprise solution.

People who would be attracted to those competitors would have never been happy with your solution anyway, and this helps your readers find the solution they need, thereby causing them to share your blog post and attract a greater readership.

How Often You Should Post

Conventional wisdom dictates that you post to social media two-three times a day, to reach all viewers. Posting too much makes you annoying and will get you unfollowed, but posting too little gets you lost in the crowd and forgotten about. With two-three posts every day, though, you can run out of things to post about easily.

Therefore, don't be afraid to repeat posts every now and again. Make sure to post duplicates or something with the same idea every 12 hours or so, to reach people in different time zones.

You can also repeat posts every few months if you can't call to mind anything to post about. Take solace in the fact that most people won't stalk your page and won't notice if you posted about the same review or promotion twice in the last six months.

Gaining Followers/Subscribers

It can feel futile when you're posting regularly, and you still have a low amount of followers/subscribers. There are easy steps you can take to gain your first meaningful followers.

A common first step is to follow a bunch of pages that are relevant to your app, in the hopes that they'll follow you back. This is called, conveniently, a follow-back. Follow-backs are not the kinds of follows that will increase sales, but they can help you increase your reach and community involvement. People also see when the amount of people your account follows are the same as the amount who follow you, which leads them to believe they're follow-backs. You should never purchase follows. Follow-backs are a good first step, but purchased follows are never a good step. A purchased subscriber will do nothing but make you look big, giving you none of the real benefits of a high number of followers.

Note Purchasing social clout is against the rules of most (if not all) social media websites and is highly frowned on. Do not purchase follows.

You can organically build audience by following the people who follow some of your app's competitors. No, you're not waiting on follow-backs to build your audience. The goal is to meaningfully engage with these people and provide value. You know they suffer from the same problem, since they follow your competitors, so if you provide more value than your competitors, they will follow you instead.

If possible, involve larger pages in your posts and tag them. If you're talking about the problem your app solves and it is relevant, tagging a much larger page in your social posts increases exposure to that page's audience as well. If they think it is good or useful information, the larger page might share the post, multiplying the amount of people it's exposed to.

One clever way to promote your Twitter account is to search for questions relevant to your app and reply to them on Twitter. It's a way you can help people right now, in the way they ask, right now. This makes your account, and you, look great. For a to-do app, this may involve searching "what's a good to-do app" and replying to those tweets. Reply to any tweets that are questions from your company account. Since these people asked for the information, it's an excellent non-spammy way to publicize your app. (Just don't fill up the feed of people who follow you on Twitter with these replies.)

Managing Your Accounts

Okay, so you know what kind of page you have and you know what kind of content you should post with it. But getting online every day at the right time on all of the social media networks sounds like an enormous pain, and it is. That is why there are services that will schedule your social media posts for you.

- *Buffer*: Buffer is a free social post scheduling service. This means that you add your social media accounts to your Buffer account, and then you queue all of your social media posts. You then program how often you want the queue to post, and it posts everything in your queue when you set it to, automatically. You only have to sit down once a week to queue up all posts for that week.
- *SproutSocial*: SproutSocial is the pricy, upscale alternative to Buffer. It does all of the wonderful things Buffer does, but also includes post effectiveness analysis, demographics analysis (for all profiles combined), and tons of other great features to make sure you're using social media the best you can.

Assessing Your Social Media

Bothering to run social media is meaningless if you're not tracking how effective certain posts are. The most effective and engaging posts should be studied and repeated, so that you have an active and effective set of social media profiles for your company. Analytics for each social media account are available through the websites themselves, as well as through Buffer and SproutSocial.

The next chapter discusses your online community as it exists across social media and email and covers how that community can be leveraged to sell your app.

CHAPTER 20

Online Community

The reason you are constructing customer support, a mailing list, social media, and a website is to cultivate an online community. An online community is a large group or groups of users who talk about your app, suggest features, and helpfully guide development. An organized community is an invaluable resource in making users happy.

Your company isn't a logo and tagline—your company is an experience for your users. Creating an online community, therefore, is about more than social media or forums; it's about getting on the Internet and helping people all over the world interact with your brand in a meaningful way. It's about those people having a good experience with your company.

An online community can be beneficial to you for a couple of reasons, discussed here.

They Can Direct Development

Your online community can offer personal input on your ideas and point you in the direction you need to make improvements. Feedback from beta-testers and users is the best thing you can have when trying to build value. This becomes useful several versions in, when you aren't sure what to update next.

Don't be afraid to let feedback from your community pivot your entire app model; sticking to a poor one because you've already put hard work into it might mean failure, when adapting to customer requests will mean success.

M. Holstein, *iPhone App Design for Entrepreneurs*,
https://doi.org/10.1007/978-1-4842-4285-8_20

It Fosters Relationships Between Users

Letting people share ideas also helps to foster a sense of participation, which will get them feeling better about your company and happier to help. That way, you have the online support and goodwill when you will need it. Additionally, people will have a richer experience if they have friends online.

They Can Beta-Test

An active and excited community will want to access pre-release copies of your app, which means you can put together a beta-testing team for free. Since they are excited and care about your app specifically, they are sure to do a thorough job and find any bugs liable to come up during normal use.

They Will Promote Your App

Since your app has proven to be a great part of their lives (providing them with a whole ecosystem around the solution they needed), community members are sure to evangelize your app to others when they need a solution to the same problem. The better your community is, the more evangelizing there will be.

Take Starbucks. App developers are always so frustrated that people won't buy an app because "one cup of Starbucks coffee costs more than most apps". But when you buy Starbucks, you aren't buying just the coffee; you're buying the entire experience of a Starbucks coffee shop.

The coffee is worth the money to Starbucks customers because it comes with a peaceful environment, music, and calm. This experience everyone gets when they go to Starbucks is the brand.

The Starbucks brand continues online, with free pick-of-the-week songs and apps, and a helpful online community (and app) with many cool things to check out. Being involved in this community gets you an impressive amount of free coffee as well, which makes participating in the Starbucks experience cheaper.

Your brand needs to create this community involvement, because everything is an experience for the person on the other end of the Internet. The experience of your app and brand is something that should keep people coming back (because it's rewarding, not because it's frustrating).

Potential and realized consumers should be able to visit your social media and get more than just product information; they should be able to get real value out of it. This can be something as small as sharing articles that people who purchase from you would be interested in, or something as major as regularly creating free content for them.

Have a Goal

Make sure to have a clear goal in mind when you are building your online community. For someone who already works another job, maximizing every minute you spend working must make a difference. You know what benefits you want from your community—input on developing your app, raving fans who will promote your app—but what does it look like for you? What kind of community actions does that translate into?

For example, people in your online community might:

- Share your app's link everywhere on Facebook or Twitter.
- Recommend your app over a work email to colleagues.
- Hop online every day to interact with the gaming community around your game.
- Help newer members of your community figure out how to integrate your app into their daily routine.
- Determine how you want users to be involved.

Make sure you're building a community where it's not just you talking at users, but where users are all talking to each other. This should be one of the goals of your community. Let users know you have forums or social media accounts where they can engage with each other.

Helpful interaction, like helpful tweets at people, is a good stating point, but you can't make that into a full community. For this to be successful, you first need people who are excited about your app. Make sure your app is worth being excited about.

How to Build an Online Community

Once you have defined your goals for your online community, you actually have to build this community. When building an online community, start with somewhere where your brand is already making people excited. If people are excited on Facebook, involve the Facebook community. If people on Instagram are Instagramming your app everywhere, get involved with the community there. Creating an online community always begins with being involved in the current one.

Tip The best place to grow your online community is with the current one.

The following section discusses techniques you can use to grow your online community beyond that initial base of users.

Answering Questions Online

One easy way to get your company name out there and interacting with people is to answer their questions online. Use the search function available in most forums and social media to search for questions in your industry, and then answer them. By getting out and offering genuine advice, you're going to be seen as a helpful company and brand. Places you can find these questions include:

- Websites on your press mailing list. Check your press mailing list for small- and medium-sized websites that have their own forums and answer questions you find there. You can also answer questions in their comment section.
- Search questions you know the answer to on Reddit and answer entries that are less than three months old. Make your answer thorough, resource-ridden, and awesome. Include a link to your app or product wherever relevant. This is so it gets upvoted and seen hundreds of times.
- Search for questions you know the answer to on Quora, which is a question-and-answer website. Highly upvoted answers can drive hundreds of thousands of people to your website and can consistently drive thousands of people a month for six to eight months. This is long-term traffic you want.

The kinds of forums you're on on the Internet can affect your image. This means not wasting time on forum websites with GUIs from the 90s or early 2000s, or websites without active users.

Find a few relevant, active sites and visit them frequently. One post here or there won't be noticed, but regular activity is always rewarded, so focus your attention on a few key websites.

Another great benefit to making your name known on the Internet and participating is that you will rise in the search engines and therefore garner more visibility. The more great responses you provide, the better your SEO.

Give Away Content

If possible, produce some secondary content that is accessible to users if they get on your social media or website. This means they have to visit your website or social media in order to download this free content. This is traditionally a tactic used to acquire emails, by providing the free content in exchange for an email, but this can also be done with pay with a tweet, if you're more concerned about social media than emails. (Which one you should be more concerned about depends on your model user's typical habits.)

For educational apps, this free content could be lesson plans or a parent-child study plan. For an emailing app, this could be a guide to getting to Inbox Zero. For this book, it came in the form of a .PDF with iPad and iPhone mockups that you can download and use for free.

You could spend hours creating an amazing product and then give it away for a couple of reasons:

- It primes your users. This is marketing-speak for saying that it introduces your app to your user, so they're more ready to download it the next time they see it. Regardless of the product, the more often someone sees it, the more inclined they are to buy. This is the logic behind you seeing the same SafeAuto ad every commercial break on TV. If they're exposed to mention of your app now, they're more likely to buy it later.
- You get something in return. You either get a share on social media, or you get an email from someone in your audience. This helps you to either get more downloads now or potentially go viral. This can be great for building a community.

- People's barriers to sharing your free product, or providing an email, are much lower when folks know they're getting a free thing in return. It gets your foot in the door, so to say.

Partnering with Other Pages

Reach Out to Competitors

The only people who really psychoanalyze company posts on Facebook are companies themselves, which in your case are your competitors. This gives you a golden opportunity to connect with those like you, to network and share what has and hasn't worked.

Reach out to them and see if other companies are up for a conversation about what they've learned being in your industry. Despite you being in competition, smaller app businesses are happy to dialogue about what works and what doesn't, therefore moving the whole industry forward. Who knows, you may find something better working together rather than competing.

Partner with the Press and Media

You can partner with the press and media to run promotions for your app. These pages may be blogs that connect readers to resources like your app, or they may be professional organizations for people in a career that would use your app. Don't be afraid to think outside the box and don't be afraid to email people with your offer.

These are the same people on your press mailing list. When you are running discounts or giveaways, these are the people you reach out to. Make sure they're sharing on their social media pages and tagging your social media pages in their posts for maximum exposure. Share any links about you they post in kind, so that your traffic is driven to their website in exchange.

Guest Post on Other Websites

One of the ways you can partner with competitors or the press is by guest posting on other company or blog websites about your topic of expertise. Subtly link to your app (or better, provide free content they can download by providing their email), but keep the focus on providing good information.

Tip Don't hide that you're the developer of your app or any content you link to, as that is unethical.

Automate Everything You Can

Once you've determined what you want your users to do, make it as easy as possible for them to do it. Find ways to automate the process as much as possible.

- If you want them to email colleagues, write the email for them and embed a mailto link.
- If you want users to share on Pinterest, provide a photo and a Pinterest share button with content already filled in.

How to do this depends entirely on what you determined the goals of your online community to be. Do research on ways you can make it as easy as possible for users to do what it is you want them to do.

Ask for What You Want

Don't be afraid to ask for the involvement from your users in the community. If you want them to share their link, add a share button and request that they share the link.

This is an especially good idea if your app has some sort of sharing mechanism, such as sharing photos or creating content. If sharing is part of your app, requesting a social share becomes even more important to your strategy.

If you want them to email you back, request so at the bottom of your newsletter.

Leverage Your Community

Once you have an online audience, it's time to get that audience working for you. If you don't interact with the audience you've built, it will wither away and become nothing. Let's go over some ways you can interact with your audience.

Guiding Feature Development

You need to be proactive in engaging with users about what they want your app to be. It doesn't occur to non-technical people that they could play a significant part in the direction of development an app takes. It's on you to communicate to them that they can.

The customer support system UserVoice does a great job of this. Recommended earlier in this book, it is a standard email and social media customer support system. But in addition to this, they have an Ideas & Feedback section of their service, where users can contribute ideas about your app and then vote on which ideas are the best. If you use UserVoice, you can list all of your different ideas on UserVoice and then simply prompt the online community to vote on which ones they like the best.

Feedback on what features users actually want becomes really useful several versions in, when you aren't sure what to update next. It can be easy to start developing feature after feature without checking whether users actually want it, and then end up with lower sales or unhappy users because you were stuffing in useless features. Just like you checked your initial app idea, always check that users want subsequent features.

Tip Don't be afraid to let an idea pivot your entire app model; sticking to a poor one because you've already put hard work into it might mean failure, when adapting to customer requests will mean success.

Beta-Testing

Reach out to your community and say that you're building a beta-testing team for future large releases of your app. Give testers the app for free. Even if you're not working on an update right now, you want to have a beta-testing team in place for when you need it, instead of wasting time putting one together later.

Use any of the beta-testing services outlined in Chapter 15 to manage your beta-testing team. Refer to that chapter when it comes time to interact with your team.

The next chapter is a case study of an app called Everest that effectively leverages communities.

CHAPTER 21

Case Study: Everest

Everest is an app that helps users achieve their dreams by breaking them into small steps that they share with other users. Users can follow one another and let them know when they have inspired them. It has all the traditional sharing vehicles, such as sharing photos, texting, and having conversations. In addition to tracking your progress, you can share captured moments with your followers.

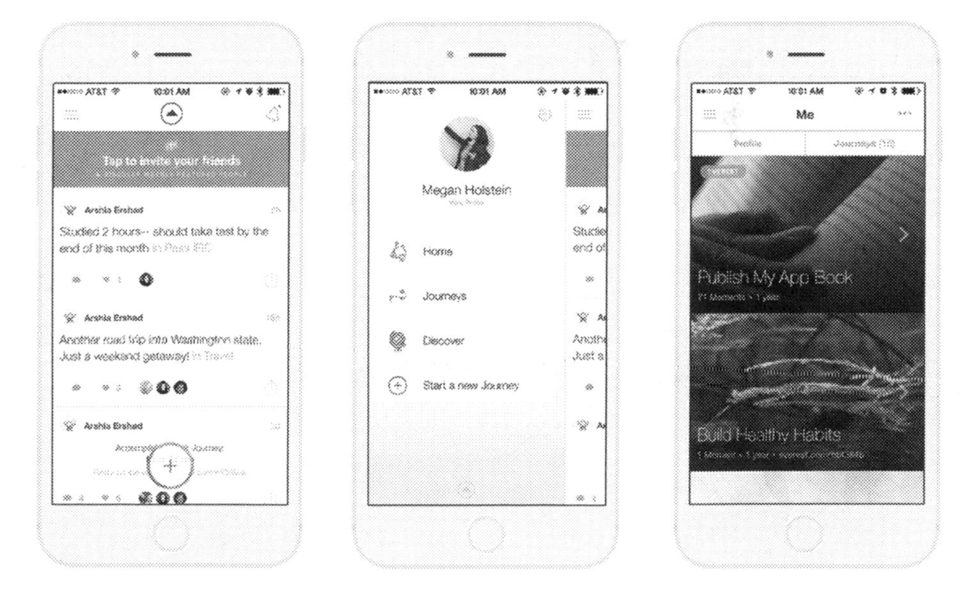

Figure 21-1. Everest app, 2016

M. Holstein, *iPhone App Design for Entrepreneurs*,
https://doi.org/10.1007/978-1-4842-4285-8_21

But what makes Everest and its team so remarkable?

Everest has received $2.5M in seed funding from visionaries like Peter Thiel. They secured funding in the center of the entrepreneurial world.

As is always the case with wild success, it took a little bit of luck. The founder, Francis Pedraza, was sitting in an airport working on app designs when a friend of Peter Thiel's saw his designs and struck up a conversation. He casually introduced him to one of the most famous venture capitalists in the world.

Don't let that fool you, though. It took Pedraza a lot of hard work to get to that airport, and then he had to keep the interest of Thiel—no small task. What others see as luck or overnight success actually has a lot of hard work and years of slogging behind it, and in Everest's case, clever branding and positioning as well. Everest has positioned themselves as an app to help people achieve their dreams, and that tugs at the heartstrings of all the dreamers out there—a lot of whom are in Silicon Valley, where Everest is based.

Everest started their journey by flying out to the valley, dedicated to networking with the best of the best. "I believe you can succeed anywhere," Pedraza begins. "However, access is power, and if you don't have access, it's very hard to get the things you need... You go to parties here, you become part of a network here. I have friends at almost any organization," he shares, because of the networking he's done in the valley. This has been an invaluable advantage for Everest and their team.

Because of this, they built a strong word of mouth for their product, which worked for them before and after their launch. "Word of mouth is different than vitality, because it doesn't happen through the product. It's not something we can just track," says Pedraza. Virality you can track, through social media analytics tools. However, virality is as important as word of mouth when speaking about app marketing. "You should be leveraging viral adoption. That's especially true if you have a free app," he says. There's such a low barrier to downloading a free app that people with free apps should be leveraging that to make a viral plan.

A viral marketing plan is any plan in which a user using the app is incentivized to get other friends to use it. With Dropbox, this is by giving people more cloud storage space for every friend they invite. With Farmville, this is done by offering people free in-game content when they invite their friends to the game. Having an app that leverages a viral plan like this can mean the difference between failure and success. Everest does this by making it easy to share your journeys with friends on your social network and to support other users on their dreams as well.

Everest was featured by Apple when they launched, and in Pedraza's words, that happened because "We had designed a beautiful product, and Apple featured it," as if it were so simple to be featured by Apple. But for Everest, it was. Getting a feature by Apple was not due to inroads or connections, but due to the fact that they built a beautiful product that people could use. You are equally capable of being featured if your app is amazing.

That said, Pedraza recommends building developer relations if at all possible. "If you're serious about being a developer in their ecosystem, it makes sense to build developer relations," he insists.

Everest's marketing plan doesn't rely on traditional methods such as ads through Facebook or traditional methods of advertising. "We ran $100 worth of tests of various channels, but we haven't done anything meaningful," Pedraza said. "It's smart to test various channels, because you never know which ones are going to work."

Despite the overwhelming success Everest has already enjoyed, Pedraza says it has not turned out like he expected. "It doesn't turn out like a rocket ship," he says. "Lots of unexpected things happen along the way. It takes four times as long and twice as much money as you think it is going to," he laughs. "Making a great app requires diligence and perseverance."

The next chapter discusses YouTube and how it can be used to sell your app.

CHAPTER 22

App Video

Online videos are one of the most popular search options in the world, and you want your app in those listings. Many people check out YouTube or Vimeo videos of apps before buying, and yet others randomly peruse videos for hours on end of things that interest them. Video is an immersive and engaging medium, and you want a video in front of these engaged users.

As with anything, a great video shows instead of tells, and your video should show everyone the benefits of your app in their lives. Let's go over how to create a video that shows the benefits of your apps to would-be users.

There are three different types of app videos you can record.

- An *informational video* informs users of the benefits of your app and walks them through using the app in its entirety. Informational videos are best for more technical apps; nobody buys a medical app because of a funny YouTube video about it, but because they see a real benefit it provides.
- A *commercial video* is more fun, bright, and emotion-evoking. The users will still see the app and learn about the benefits. They inspire users to download the app, as opposed to providing information about it. Nobody buys a game because a video informatively told them, "Playing flim flam will increase entertainment and happiness in your life." That app video would have to show the viewer that the game will evoke some sort of emotion.

M. Holstein, *iPhone App Design for Entrepreneurs*,
https://doi.org/10.1007/978-1-4842-4285-8_22

- The *company video* is appropriate for people who are employing the long tail model and have a suite of apps with the same target user. It's not a video promoting all of your apps, but a video that familiarizes the public with your company. Who are you and what do you stand for?

Informational Videos

Informational videos walk the viewers through your app in a more intellectual and clinical matter. You still prioritize benefits over features, but the benefits are more conceptual and intellectual in nature (as opposed to the emotional pull of an iOS game).

Informational videos should be in a neutral setting, like a plain desk or simple office. Neutral does not mean unattractive, so always make sure it's bright, nice-looking, and appealing. Keeping it visually simple will allow users to focus on what the iDevice screen is doing and on learning on how to use the app.

This is a video type that uses screen recordings of the iOS Simulator with someone explaining features, rather than someone awkwardly holding the device. A visual of you can be in one panel, walking the users through your app, and then the actual device can be on the other screen.

This approach is recommended if you can't get the contrast of your setting to work right, as you don't want your video to show your hands as big black shapes. Don't be afraid to talk to the camera like you would a person, as your listener is on the other side.

Tip Use real audio and speech, not subtitles. Using only subtitles in videos isn't wise, as it makes it difficult to follow the information being presented. It also conveys cheaper production quality.

Promotional Videos

The purpose of a promotional video is to paint a picture of your users. This should be the person the viewer wants themselves to be or sees themselves as. This builds the association that using your app will make them like that.

Alternatively, this could be the sort of emotional video that, when it's over, people take a moment to sit back in their chair and enjoy the inspirational or moving emotions they felt. Piano or violin music, beautiful camera shots, and attractive people or products will surely play a part in these videos.

In your video, try to avoid gimmicky phrases like "Unlike other apps," "What we do different," and so on. It should be fun and engaging, but it shouldn't be as bad as an infomercial. Go over your script to make sure it sounds engaging.

Don't be afraid to present yourself as the only reasonable option. This isn't an informational video, and you are trying to wrap your viewer into the emotional sell of your whole app experience. Let the viewer be immersed.

Present the unique selling points of your app up front. If your app is a task manager, don't tell them about the things your app does that any task manager does, tell them about the thing *only yours* does.

Company Videos

The purpose of a company video is to introduce the viewer to the company's core goals and values and to demonstrate how the app line your company produces contributes to achieving those goals. This gets the user emotionally invested in your goals, and then they're presented with a way to achieve them.

A company video is not a video promoting all of your apps, but a video that familiarizes the public with your company. Who are you and what do you stand for? When people are looking at your company possibly for business or to feature your product, the consideration will make you stand out.

This video type would also be especially good for those with expensive apps, or those who are providing apps to a concerned/dedicated segment of the public or on a controversial topic. The video should take a tour of the workspace and have a conversation with everyone on the team if the team is small. If the "team" is just you, style it like an interview.

Tip Make sure to cover the mission and vision—why you are doing what you're doing, why you started, and who you mean to help.

Make sure to keep the video interesting; instead of just looking at the camera and droning on about your company, provide visually interesting and beautiful camera shots and inspirational instrumental music. This adds an emotional element to your company as you're speaking about what you do. Consider some stock footage of your app in use, or you working.

Recording Your Video

You can record your video in one of two ways. The option that works best for you depends on the type of video you are going for. Let's go over how to record the two different types of app videos.

Video Recording

Once you've decided what kind of video you want to have and have laid out the basic progression of your video, you can hit the recording equipment. To do so, you need to know the basics of capturing an iDevice on film.

As with any technology, it can come out looking pretty strange in the film if you're not careful. The next sections discuss a couple of guidelines to help you film on an iPhone.

Keep Your Video Shorter Than 30 Seconds

People's enthusiasm will begin to wane after the first few seconds of any video, so you should put all of your engaging and short material in the front, to keep viewers watching it. Remember to keep your video fun, as well. If you have fun, it will show.

Make the Focus of the Photograph Where You Want It to Be

Having different things in focus can affect the subject of your shot. You can choose to have something else in focus with the app, or just the app itself to communicate different things to viewers. An example of this is provided in Figure 22-1.

Figure 22-1. *Distant versus close focus*

Make Sure the Exposure Is Correct

One common problem is overexposing/underexposing the screen (as seen in Figure 22-2), so that the surrounding environment isn't very attractive. You're going to have to fidget with the brightness on your device and the exposure setting on your camera. The automatic adjustments (on either device) probably won't do you much good here, so you should turn them off and manually fidget with your brightness settings.

Figure 22-2. *Well lit versus poorly lit*

Pick a Good Environment

The environment in which you shoot your video should not have very much contrast, as your camera will white-out parts of the shoot if it does; common cameras are not as good as the human eyesight (yet).

Make sure any lighting for your video is natural, as opposed to yellowed artificial lights. It's not that the shot has to have a "natural" feel, but natural light brings out bright colors whereas artificial light only yellows them.

Screen Recording

Instead of hassling with recording a device, you could fire up your iOS Simulator in Xcode and do a screen recording of that instead. If you're handy with video editing, you could pair up this screen recording with a video recording of you alongside your app on the computer, walking viewers through your app.

To do your screen recording, you can use QuickTime. QuickTime Player comes on all Mac computers and can be upgraded to Pro for a small fee. QuickTime records your screen and audio, although not at the same time, so you will have to be handy with video editing software to get the audio and video to sync up. When you do your screen recording, make sure your computer desktop is clean and shows your desktop image.

Editing Your Video

On Apple MacBooks, there are two good options produced by Apple for video editing software, which are variations of each other.

- *iMovie*: Apple has already taken the time to develop consumer video editing software, called iMovie. It is available in the Apple Creativity Suite, which comes with iPhoto, iMovie, and GarageBand, which all MacBooks come with. This is their complete suite of consumer photo, audio, and video editing software, and is more than enough for you to put together your app video.
- *Final Cut Pro X*: Final Cut Pro is the fancy, expensive version of iMovie. It is for professional videographers and has a lot more functionality in terms of splicing audio and video. It results in higher production quality. This software will cost you a pretty penny.

The next chapter discusses how to get your app in the news.

CHAPTER 23

The Press

Everybody wonders what the big secret is to getting picked up by a major news outlet. The problem is that the big secret is to be famous. So how do you generate buzz if you're not already famous?

One of the best ways to get covered is to contact press outlets and blogs when your app comes out. You can find press outlets to reach out to on the mailing list you built earlier, as well as by searching the Internet to find new ones to add to it.

In addition to emailing news outlets, you can tweet at them from their corporate twitter accounts. People are often more responsive on Twitter than they are through other channels, and on Twitter you're reaching out personally, not just to a contact@us.com email.

Don't assume the big guys won't take notice of you. The worst they could do is say no, so go out on a limb and contact major outlets too.

Note A notable example of this is Steve Jobs, who phoned up the president of Hewlett-Packard like it was nothing and asked for some computer parts to mess around with. Bill Hewlett liked Jobs' spunk, and then right after he graduated Jobs went to work at Hewlett-Packard, having developed a relationship with that corporation.

M. Holstein, *iPhone App Design for Entrepreneurs*,
https://doi.org/10.1007/978-1-4842-4285-8_23

A lot of people who have been successful were so because they were willing to go out on a limb and contact the big names. The absolute worst anyone could ever do is ignore you, so go out there and contact someone important.

Even the most important and busy people are still people. They have likes, dislikes, dreams, favorite restaurants, and a sense of humor. Humanizing someone makes them a lot easier to approach, so approach everyone you can, keeping in mind that they put on their pants one leg at a time too.

Tips for Contacting the Press

Personalize your emails. Personalized emails are the best way to get the attention of the press. Do not generate an emailing list and send everyone a stereotypical, cheesy email. Read over each website, learn the name of the person on the other end of the emailing address, and send them something that proves you took a look at the real website and that this is a personalized email (perhaps comment on an earlier article or blog post they wrote within the email). If you want to go above and beyond, go to their website and comment on a recent article they've written in the comments section.

Offer to let them use your app. You can use a beta-testing service mentioned earlier to get these copies of the apps to them early, or use iTunes promotion codes to get them free copies after your app is out. This allows them to get a review copy for free, which before release is a privilege.

Be available to contact. Make it obvious how to get in touch with you, and that you are available. Provide phone numbers, emails, Twitter handles, everything you have that allows them to reach out to you. This lets the people on the other end know they're important, and that you think their time is valuable. Besides, you wouldn't want to miss an inquiry from the press.

Tell a story. When emailing people about your app, don't open up with "Hey, I made this app called Puzzles; will you review it?" Tell them a little about yourself, your app, and why you made your app. This is especially true if your app has a compelling story behind it. Even if you're just a guy who wanted to have a go at making a game, you have a story. Maybe you're a go-getter husband-and-wife team, someone trying to put themselves through college, or you're a parent trying to assist your child. There is always a story, and media outlets don't run on apps, they run on stories that interest their readers.

People respond to news a lot more when there's a story that they can enjoy. For example, if your press says that you made a math app to help your cousin do better in math class, it will garner a lot more sales because readers will purchase the app to help their kids do better in math class as well.

Include an example of your app in the email. This can be a couple of screenshots or a video of your app in action. Allow the reporter to get enrolled in your app vision and more likely to write a piece about your app and your story. This saves them the effort of downloading it themselves. If they can see it in person, they're much more likely to write a story on your app, because it's more real to them. If you have a lot of materials you could share, consider attaching a press kit to the email.

Don't attach these directly to the email, but host them on a service like MediaFire or Dropbox, so that the attachments will not by rejected by the other end's mailing service. Some corporate servers reject emails with attachments as spam and route them directly to the spam box. This will also make the email a lot easier to load on mobile.

Believe in yourself. Don't sound timid when you're emailing anyone about your app; truly believe in your story. No matter how badly worded emails can be, true enthusiasm and sincerity about what you're doing always shows. People pick up on this, and it can be a make-it or break-it difference. Also, there's going to be enough people who insult you or your app, and you don't need to be one of them.

Not only are these tips for contacting people about your app, but they are tips on how to present yourself to anyone. Show enthusiasm whenever telling someone about your idea and don't let people full of doubts tell you you won't make it big. Especially on the App Store, you can't win the lottery if you don't even purchase a ticket.

What's nice about the App Store is that it is more than gambling; a better app will get you a lot closer to winning, and it isn't quite as based on luck as you would think. The millionaires of the App Store struck it big, but skill can carry you further than $5,000-$10,000 a month on one app.

After that, people will pick up on your idea and run with it, and you will be dragged along for the ride—that is how app million-dollar businesses like Rovio were created.

Write a Press Release

If it's a larger and more official outlet you're reaching out to, you may want to use a press release. A press release is official documentation of news that you send to the press. It is commonly used with older and established major media outlets. It is, in effect, doing all the fact-finding and quote-getting on behalf of the outlet, so that they can get directly to writing up and publishing the news.

Reaching these media outlets can lead to all sorts of exposure, and that's where the press release comes in. A well-written press release can impact your sales in a significant way, especially since stories can be picked up from one outlet and reported on in another. You can also send press releases to actual PR websites and firms, who will distribute it on your behalf.

Press releases are an older form of distributing news. Because of that they have formatting mechanisms that, although archaic, are still respected today. When writing a press release, always write in a formal style with active voice, from a third-party point of view. Active voice places the subject as the first noun in the sentence, like so:

- Active voice: The hunter saw the deer.
- Passive: The deer was seen by the hunter.

They always say, along the top left, FOR IMMEDIATE RELEASE or HOLD FOR RELEASE UNTIL... to indicate whether the story should be run ASAP or held until a certain time. They also always end with ENDS or ### to indicate the press release has ended. You can also optionally include the word count of the press release. An example of a press release and its elements are included for you:

> FOR IMMEDIATE RELEASE:
>
> Game Puzzles to Sweep App Store
>
> Hamilton, New Zealand - November 12, 2012 - The app Puzzles is next to sweep the App Store, rising to #15 in the charts in just two days.
>
> Released on November 10th by Cool Games Ltd, Puzzles is a new board game designed specifically for the iPad, a spin on *Chutes and Ladders.*
>
> Cool Games Ltd. decided to make Puzzles as a child friendly board game for those who would rather keep an app around than a board game.
>
> Puzzles was inspired by Katie, daughter of Shelly and Marcus Smith, the owners and two-man team that makes up Cool Games Ltd. Katie doesn't play board games often because of the cleanup, but she does play games on her iPad.

Contact:

Shelly Smith

email@example.com

233 North Street,

Hamilton,

New Zealand

Ph: +64-877-9233

###

A press release is not an ad, and it shouldn't be worded like an ad. A press release should be useful, interesting, and accurate, interesting being the most important. Your press release may be useful and accurate, but if it's as dry as a law textbook, nobody is going to report on it. The press release has to present your app as something newsworthy.

Simply existing isn't enough for your app to be newsworthy, so time a press release around update or release date. Furthermore, to make your press release interesting, don't talk about the app features, or even benefits.

Facts and statistics are something people love. Quantifiable, quote-able things capture people's attention. Be sure to use relevant facts and statistics in your press release, such as mathematics literacy rates and how your app wild improve.

Make sure to keep your press release concise. Tell readers everything they need in the first paragraph and don't drag out their concentration. Reporters often receive up to dozens of press releases a day, and wasting their time is going to turn them off of publishing your piece.

A good way to keep a piece concise is to avoid the word "I". People often don't realize how much we're talking about ourselves until we make a conscious effort to rid that one word from our vocabulary. Not only is it a useful trick in writing a press release, eliminating the word "I" from everyday vocabulary is a goal for many people.

The most common press release people use for apps is prMac, although there are a few others people use. This free press release method is a great idea for proof-of-concept for a small app, before you take the time to expand. However, you needn't use a press release service and can just send your press release to your media list instead. Always make an effort to add websites to your media list before sending a press release.

Infographics

Something people share often that you could make is an infographic (as seen in Figure 23-1). An infographic is a very attractive graphical representation of data, with a focus on facts, statistics, and emphasizing numerical conclusions. Infographics have high publish rates because they're visual, making them easy to understand, and they're highly relevant, meaning the facts and statistics apply to the reader's life.

***Figure 23-1.** Sample infographic about the United States*

The caveat of infographics is that they need numbers, charts, and visual representations, which can cost you to hire a graphic designer to produce. There are services online for creating and promoting infographics, to avoid the hefty charge that comes along with commissioning an infographic designer. One service like this is Piktochart, which you can use to create your own infographic for free.

If you don't want to take the time to make your own infographic, you could include one that already exists in the press release or promote one on your social media. You risk directing people to the infographic creator's website instead of yours this way, but it could still put your app out there to be seen, and when shared, would still link back to your page. It saves you time and money that you'd have to spend creating your own.

Press Kits

You'll want to take the time before contacting the media to prepare a press kit. A *press kit* is a bundle of logos, app screenshots, and other information someone needs to do a write-up about your app. They are often found as a .zip file on company/app websites, or they're attached along with press kits and press inquiries in emails.

Include one when making your press inquiries. Don't make reporters look for screenshots, details, or contact information; have it all zipped up right in the email for them to unload. Additionally, by pre-making it, you know it will look polished and impressive to a reporter.

However, you don't want to attach it directly to the email. Some corporate emails will bounce attachments from foreign servers. To avoid emails being bounced because of attached materials, upload them to a service like MediaFire146 or DropBox147 and then copy that download link into your emails. Additionally, you'll be able to see how many times your press kit was downloaded, telling you how many reporters perused your email further.

Additionally, check your press release in plaintext to see if it looks okay before sending it. Not all people view their emails in HTML, and you want to make sure your email looks fine in plaintext before sending it to the press. Similarly, some corporate servers will only display emails in plaintext initially, so your email still has to be engaging and eye-catching in plaintext.

The next chapter discusses buying ads, including when it is or isn't appropriate for your app.

CHAPTER 24

Advertising Space

Should you invest in ads for your app startup? If the answer were a clear and concise "no," you wouldn't ever see an app advertised, but you do. But the answer is not a simple "yes," either.

Many companies do buy ad space for their apps. We've all seen a commercial for a service that ended with a request for you to download their free app, or a banner ad in Google ads for a new free app.

The thing is, they're all companies that do something beside make apps. AutoTrader runs ads for their app, but AutoTrader is actually a whole car-trading platform. AirBnB runs ads for their app, but they are a full-fledged rental lodging website. Travelocity runs ads for their app, but they're a complete online travel agency. None of these are standalone apps, but companies with apps.

What you don't see are ad campaigns for individual apps. Scatter-bombing ad campaigns work for large companies because they have different goals for their campaign than you, the indie developer, does. The goal of an indie developer is to get the most bang for your buck, meaning every dollar spent on advertising should get you the most downloads possible. You don't have enough money to reach the whole market, so your goal should be to get the most downloads for your money.

Large companies do have enough money to reach everyone. Their goal is to reach the whole potential market, so that when the download does occur, it occurs for them. It's a long-term strategy that requires a lot of money. This is why you see so many ads for insurance, car companies, and

M. Holstein, *iPhone App Design for Entrepreneurs*,
https://doi.org/10.1007/978-1-4842-4285-8_24

other large purchases—so that when it does come time to purchase, you go with them. It's not a good strategy for small, emotional purchases.

With millions of apps on the App Store, a scatter-bombing marketing campaign with Google banner ads or keywords won't give your app the exposure it needs to stand out on its own. People don't like looking at ads or clicking on them, and even if something catches their eye and they do click, they probably won't buy. Less than 5% of people who see an ad go on to immediately buy the product/service.

Even if you did get the right exposure from these ad campaigns, it's just too expensive. The average per-click advertisement campaign runs about $2 or so per click (across all advertising platforms), with a conversion rate of 5% or less. Unless you're charging $9.99 or more for your app, you're not going to make a profit this way.

Purchasing advertising space can work, but your plan has to be well thought-out. You can't purchase advertising space on *MacLife* for your app, because that audience is not targeted enough. You need to target a highly specific market to have the greatest odds of that exposure turning into downloads.

Refer back to the specific target user you identified in the first section of this book. You identified age, gender, education level, income, life circumstances, and lots of other characteristics in order to create your model target user. These are the people you want to advertise to, and only them, in order to get the most downloads.

Tip If you've installed the app analytics and website analytics mentioned earlier, as well as checked your social media analytics, you should have a pretty good idea who your users have been so far. Use this information to adjust the model target user you made in the first section of the book.

This seems counterintuitive, because at first thought you want as many people to see your message as possible so that more people download. However, the choice between more or fewer people seeing ads is usually priced per click or per impression, so you'll get as many impressions/clicks as you pay for. You want to make sure that money per impression is spent on a viewer who is most likely to turn into a sale—the target user.

The cheapest way to get ad space in front of your target user is to purchase banner ad space on the niche websites they visit. Small websites sometimes manage their own ad space, and due to their inability to track impressions and clicks as well, will instead charge a flat fee per ad. This can be an excellent advertisement deal, since it's a cheap ad that targets your model target user exclusively.

Note You identified these niche blogs and websites while building your blog/press mailing list earlier in this section.

Next, you need to do a bit of math to determine if your ad is profitable. If you just purchase ads and pray, it's easy to end up losing money on advertising. When you purchase an ad, track how many *additional* app purchases you get over your *average* that month. If you make more money with *additional* purchases than you spent on the ad, it was profitable.

Designing Your Ad

If you're going to buy an ad, you want to make sure it's designed well, or you're wasting your money. The following sections describe the things all advertisements should have.

A Call to Action

Make sure that your ad says "Available on the App Store" to clearly communicate that it is an iOS app, and that they can click the ad to download your app. If possible, use the image Apple provided.

This is a call to action for the viewers that they should download your app upon seeing the advertisement, and it works because it relies on symbolic associations that Apple has already built for you.

Good Design

If you're not very good at graphic design, hire someone who is. The attractiveness of the ad is what catches people's eyes in the first place, so you want to make sure it's visually appealing. Compare your design to other great designs on the Internet—keep your eye out for advertisements you like and emulate those.

Minimal Text

Viewers will be looking at your ad for only a second, so don't put any more text than absolutely necessary on the ad. A slogan or other longer sentences will be passed over, unread by the viewers, and can actually cause your ad to perform worse. Match the amount of text on your ad to the time the viewer has available to read it.

Keep in mind that purchases as small as apps tend to be emotional in nature, not thought-out. Attractive design and simple emotions will motivate your users to purchase.

The Price (Especially If It's Free)

A free app has no barriers to download, and people will be much happier to download a free app on a whim. If they see a high price, the viewer might cringe and move on. Get them more excited about your product before displaying the price if it is high.

A good example of advertisements are Apple's, which sometimes stop being advertisements and start being art. They tell you nothing about Apple's products, but somehow still make them look like the upper-crust tech accessories everybody wants (even if that's not necessarily the case).

Here is an iPad mini billboard ad by Apple (see Figure 24-1). This ad doesn't even say anything about the iPad mini. The product itself takes up less than half the space of the actual advertisement. If this wasn't a good thing to do, someone at Apple would have halted the release of this series of advertisements, yet these modern and empty advertisements peppered the skyline of every major city. These ads work because they make the mini look cool. We don't buy the mini because it has a whatever-gigahertz processor—we buy it because of the emotional feeling we get from having one.

Figure 24-1. *iPad Mini billboard, 2016*

Print Media

Unless you have a compelling reason to buy advertisements in print or physical media, do not. The advantage of apps is they can be purchased online, anywhere, on the slightest of whims. This advantage is forfeited with print media.

With digital advertisements, someone can click on your ad and buy your app several seconds later. With print, the viewer has to take note of the app, remember it, look it up later, and then maybe buy it.

We can't remember our own grocery lists in this day and age, let alone an app we saw on a billboard for several seconds on our drive home. Even the most motivated consumer will sometimes forget the app they saw, dropping your conversion rate from an already-low 1% or 2%. Your money is better spent elsewhere.

The next chapter, which is the last, discusses how to use discounts and free promotions in order to sell more of your app.

CHAPTER 25

Discounts and Promotions

Once your app is out, you can periodically run standard discount campaigns and promotions to boost sales of your app. These are quick and easy things you can do periodically to keep sales up and your app relevant.

Because you have spent the time building an infrastructure for your business (social media accounts, a press mailing list, and an established business model), executing most of these discount and promotion techniques takes no more than an hour or two of your time.

Standard Discount Campaign

The most obvious discount or promotion campaign is to actually discount the price of your app for a certain amount of time. This is essentially the same as a free-for-a-day campaign, except the price is only partially discounted.

Partial discounts are only effective with apps that are expensive and have a lot of leeway for discounts, such as apps more expensive than $4.99. Discounting an app from $1.99 to $0.99, while a 50% price reduction, isn't usually a large enough amount of savings to get users really excited.

M. Holstein, *iPhone App Design for Entrepreneurs*,
https://doi.org/10.1007/978-1-4842-4285-8_25

How to Discount Your App

- Go online to iTunes Connect and pick a date to set your app as discounted that is two or three months in the future. This gives you adequate time to set up marketing campaigns with various online outlets.
- Write up and send an email or press release. Make sure to mention the size of the discount, the new and old price of your app, and the duration of the discount. If there is a holiday or occasion for the discount, mention that too. Politely request that they share the discount with their readers on their website, mailing list, and social media.
- Schedule a few posts on your own social media in the week leading up to the discount. This way, anyone who is interested can share the posts as they wish.

Free For a Day

A great way to drive sales for your app is to make it free for a day. Like any other discount or promotion, if you prepare and publicize it correctly, it can result in a large sales increase. Your free app gets publicized the day of the promotion. People will click on the link in the following days too, not realizing that the app isn't free anymore, and many will decide to buy it anyway. Having a ton of free downloads pushes your app up in the charts.

At first, it may seem counterproductive—surely the people who are downloading the free app might have bought it instead. Aren't you losing money by doing this? In fact, not really. The majority of people who download your app while it's free wouldn't have paid for it. For the people who would have paid money, there are always more people out there willing to pay money too. The App Store is extremely large—over 75 billion apps have been downloaded—and that means there's quite enough people for you to give your app away to a few of them and still have significant revenue.

This tactic is best employed when you're already connected to the community leaders of your users. This means promoting through your social media or website, as well as through the leaders of other blogs, forums, and websites where your users hang out. We discussed putting together a mailing list of related blogs and websites to contact with news; now is the time to use it.

How to Run Your Free-For-a-Day Promotion

- Go online to iTunes Connect and pick a date to set your app as free that is two or three months in the future. This gives you adequate time to set up marketing campaigns with various online outlets and gives the notoriously unreliable iTunes time to update. Remember to set both start and end dates. You must schedule it to be free and then schedule it to be paid again the next day.
- Write up and send an email or press release (depending on the tone of the outlets in your press list). Mention the scheduled date of the free day and make it clear that readers would be interested in this free stuff. Most importantly, make sure your outlets know there are no strings attached. Request that they share the free-for-a-day promotion with their readers on their website, mailing list, and social media, even if they don't want to do a whole write-up on it.

- Schedule a few posts on your own social media in the week leading up to the free day, letting people know that the app will indeed be free. This way, any other people who have websites/blogs/followings can share the information as they wish.
- Schedule emails to your opt-in email list, letting them know it will be free as well. Send them only one or two emails; they don't need to be bothered as much as folks on social media do.

This should take an hour or two of preparation. If you contacted enough media outlets of the appropriate sizes (small to medium work best), you will see satisfactory payoff with your promotions.

This tactic is effective if it's used infrequently, but if you do it too much, it loses its effectiveness. Outlets cease to see it as a big deal, and viewers will wait to buy if they know it will just be free again later.

Promotional Codes

Another way you can promote your app is by giving away promotional codes. Apple gives each app version 50 promotional codes, and you can use these to get free press and exposure for your app. Apple intends these to be given away to press outlets that can give your app more exposure. This is based on a practice in the software industry at large, where pre-release and free copies of software are given to the press.

This means that you give away promotional codes to websites that have the same audience as your app, and they will review your app with the free codes and give the rest away to their readership. You get free exposure, and their readers get free stuff.

You can increase your return from these promotional code giveaways by having users receive more chances at winning for doing different things. For instance, readers may have to like the company's Facebook page and/or Twitter, and maybe their Pinterest or other social network. Each additional hoop gains people more entries. These hoops should reflect your business goals. This can easily be done by requesting the website administer their giveaway with a service like Strutta, Rafflecopter, or Kontest.

The advantage to this promotional method is how easy it is—10 minutes of emails. Five minutes to make contact with the site and provide the codes, and five minutes to read the confirmation email and the review itself. This is so easy because you already took the time to build a press list.

Smaller websites especially are willing to take the time to honestly go over your app with you and determine how beneficial it is for their readers, so that when you do approach larger outlets, your app is better and more worth their time.

Places to Give Away Codes

Consider these places to give away promotional codes.

Niche Websites

These should be the same niche websites you identified for your press list and to advertise on. Each blog or website does a piece on your app and gives away promotional codes at the end, and you get free press and a free detailed review in exchange for three or four promotional codes per website. The wait time on this review will vary from website to website.

Dedicated Facebook Communities

Many app markets—such as educational, action games, productivity, and other active areas—have dedicated Facebook communities that consist of the Facebook pages of these websites. You can reach out to these social media communities from your social media page and offer them promotional codes to give away in exchange for a review.

Running Your Own Giveaway

You can run your own contest, but be careful of Facebook and Twitter's promotional policies, as they dictate a couple of things. Among them is the need to work with a third-party application to administer the giveaway, such as Strutta, Rafflecopter, or Kontest. You can publicize your giveaway or any giveaway on your Facebook page or Twitter, but entries need to be counted somewhere that isn't a social media feature, like a Facebook Like, a Facebook comment or a Twitter follow, per the guidelines of the websites.

Tip Always remember that if a giveaway is going on anywhere, you can post about it on your social media websites to keep users engaged.

Some websites will offer you paid reviews. They're the same thing as a free review, minus the wait time. A paid review never guarantees a positive review, it just guarantees a speedy one and prominent exposure of the review on their website. Paid reviews will come with the purchase of advertising space, or you usually can pay a small fee to the website. The allure of this is that the turnaround for a free review can be months and months, but a paid review is always returned in a matter of days, since a paid review is always a priority in the website's eyes.

Blogs will usually review your app with the expectation that you will cite the review and possibly quote it on your website or in your app description. This doesn't mean they always will let you, though; be sure to ask permission to do so. Often, these posts will be careful not to leave any quotable lines that you can use for your iTunes description as well. This is great for the reader, but not for you—you've essentially thrown away those promotional codes. This is bad because you can't get promotional codes back—once you're out, you're out.

You can request the winners of the giveaways leave a review of your app as a courtesy. You could have the website state that a review is expected in exchange for a promotional code, although it cannot be mandatory for winners.

It is against iTunes Connect's contractual agreement to force people to leave reviews at all, regardless of what the review says. Breaking this agreement can get you banned from iTunes.

This can be irrelevant though. Unfortunately, not all blogs that run giveaways actually review the apps they're giving away. Their post might include a few lines, such as "Doodle Jump, a fun and addictive game for people of any age." But nothing expansive. Discuss review expectations individually with each website, especially if you're hoping for a review.

APPENDIX

In Closing

There's a lot more to making an iOS app than what was contained in this book. This was just a primer to all of the various disciplines involved in creating an app, and any single chapter in this book is in reality its own job description and career path. What you've learned in this book is enough to begin making apps, but don't let your education stop here. Go pick up more specialized books, books about programming or marketing or design. There is much more to learn.

When You're Done Reading

If you have time, please leave a review on Amazon. Authors need reviews to know how they're doing, so your review is critical.

If this book has helped you, there's a lot more where that came from. See the author's additional work at meganeholstein.com.

M. Holstein, *iPhone App Design for Entrepreneurs*,
https://doi.org/10.1007/978-1-4842-4285-8

Index

A

B

M. Holstein, *iPhone App Design for Entrepreneurs*,
https://doi.org/10.1007/978-1-4842-4285-8

L

M, N

O

P

Q, R

S

T

U

V

W

X, Y

Z

Printed in the United States
By Bookmasters